Electrical Installation Technology 3

by the same author

Electrical Installation Technology 1
Electrical Installation Technology 2
Multiple Choice Questions in Electrical Installation Work
Worked Examples in Electrical Installation

Electrical Installation Technology 3

Maurice L. Lewis
BEd (Hons), MIElecIE
Course 236 Tutor in Electrical Installation Work
at Luton College of Higher Education

Hutchinson
London Melbourne Sydney Auckland Johannesburg

Hutchinson Education

An imprint of Century Hutchinson Limited

62-65 Chandos Place, London WC2N 4NW

Century Hutchinson Australia Pty Ltd
PO Box 496, 16-22 Church Street, Hawthorn, Victoria 3122, Australia

Century Hutchinson New Zealand Limited
PO Box 40-086, Glenfield, Auckland 10, New Zealand

Century Hutchinson South Africa (Pty) Ltd
PO Box 337, Bergvlei 2012, South Africa

First published 1984
Reprinted with amendments 1985
Reprinted 1987

Set in IBM Press Roman by Tek-Art Ltd, London SE20

Printed and bound in Great Britain by
Anchor Brendon Ltd, Tiptree, Essex

British Library Cataloguing in Publication Data

Lewis, M.L.
   Electrical installation technology 3.
   1. Electrical engineering
   I. Title
   621.3   TK145

ISBN 0 09 152611 6

Contents

Preface

Acknowledgements

This is the final book of three volumes intended to meet the study requirements of electrical installation work students pursuing the City and Guilds of London Institute Course 236.

The aim of this book is to cover topics pertaining to the Course 'C' syllabus and it will prove a useful textbook for Course 232 and TEC students since the chosen chapters focus on requirements for safety, supply and distribution of electricity, special installations, lighting and heating, planning and estimating, and electronics.

It is worth mentioning to Course C students that possession of the CGLI Course 236 C Certificate along with both Part II and Electrician's Certificates will qualify them for the award of the City and Guilds Licentiateship (LCG). This award is in recognition of achievement in education, training and employment.

Electrical installation work courses

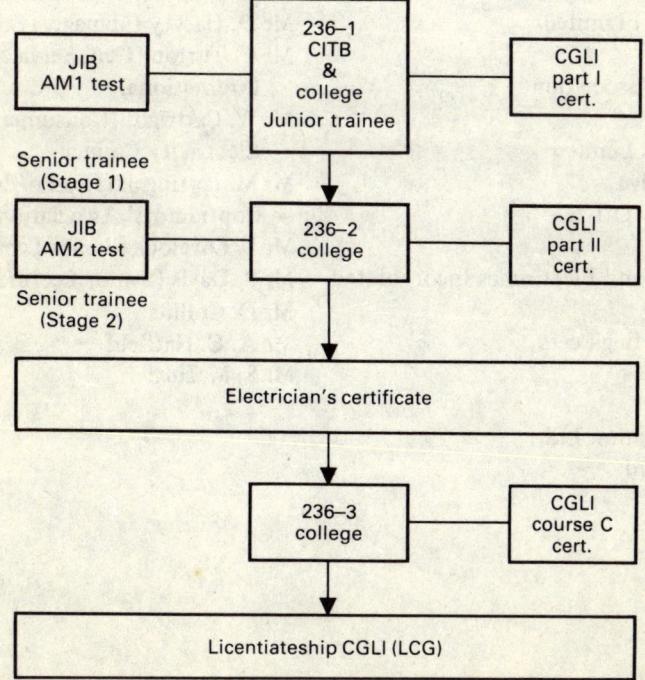

Acknowledgements

The author wishes to thank the following manufacturers and organizations who have contributed information to make this book possible:

Agricultural Training Board
APC International
Avery-Hardoll Ltd
Barton Conduits Ltd
British Standards Institution
Central Electricity Generating Board
Chevron Petroleum (UK) Limited
Chloride Gent Limited
Electrical Contractors' Association
Electricity Council
GEC Industrial Controls Limited
Health & Safety Executive
Her Majesty's Stationery Office
Impalloy Ltd
Institution of Electrical and Electronics Incorporated
Engineers
Institution of Electrical Engineers
JSB Electrical Ltd
KAC Alarm Co. Ltd
Lighting Industry Federation Ltd
Midland Electricity Board
Solus Ocean Systems

Also the City & Guilds of London Institute for allowing past examination questions to be reproduced.

The author also wishes to take this opportunity to thank the editorial staff of Hutchinson Education for making the series possible and also the following people who provided information and help for Volume 3:

Mr C. Short (Contracts Manager) Fine & Humfrey Ltd
Mr J. Lerche (Partner) APC International
Mr D. Harvey (Manager) Solus Ocean Systems
Mr C. Turton (Gen. Services Section) Midland Bank International
Mr W. Cartright (Consumers Technical Engineer) The Electricity Council
Mr M. Pettingell (Head of Information) The Electrical Contractors' Association
Mr T. Lovelock (Senior Lecturer) Luton College of HE
Mr S. Davis (Senior Lecturer) Luton College of HE
Mr D. Collins
Mr A. C. Hatfield
Mr S. McHugh

chapter one

Requirements for safety

After reading this chapter you will be able to:

1 State a number of safety requirements concerning the Electricity Supply Regulations 1937.

2 Describe briefly the important chapters associated with the IEE Wiring Regulations for Electrical Installations 1981.

3 State the purpose of British Standards and know the use of BSI kite-mark and safety-mark.

4 State a number of additional requirements for safety, particularly the involvement of BASEEFA in testing and certifying equipment in flammable atmospheres.

Some requirement aspects of both the Health and Safety at Work etc. Act 1974 and the Electricity (Factories Act) Special Regulations 1908 and 1944 appear in Volume 2 of this series, and those students who have read this book and/or passed the City and Guilds Course 236 Part II examination should now be capable of demonstrating their knowledge of these two documents in more practical installation application. This particular chapter sets out to complete the requirements for safety with regard to electrical installations, concentrating on topics from the Electricity Supply Regulations 1937 and the IEE Wiring Regulations for Electrical Installations 1981. A brief mention will be made concerning British Standards as well as additional safety requirements which are expanded upon in later chapters.

Electricity Supply Regulations 1937

These are the mandatory regulations concerned with electric lines and works of the supply undertakers (supply authorities) as well as the supply of energy to consumers' installations. A summary of Regulations 22–29 and 32 now follows:

Regulation 22 is concerned with the service lines into consumers premises. It relates to underground supplies where there may be a need to prevent any influx of gas at the point of entry into the premises.

Regulation 23 is concerned with the identification of the supply lines or cables by means of colouration or labelling etc. This is really to indicate the polarity of the live (phase) and neutral conductors at the supply terminals.

Regulation 24 deals with protection of consumers' installations against excess energy, either by using a suitable fusible cut-out or by automatic circuit breaker of adequate rupturing capacity which must be completely enclosed in a suitable locked or sealed receptacle of fire-resisting construction. The regulation forbids any fuse or circuit breaker in the earth conductor, and where the supply is at high voltage, provision has to be made to isolate the fuse or circuit breaker from the service line, thus allowing the supply voltage to be cut off at the intake position.

Regulation 25 gives the undertakers the responsibility for all electric lines and apparatus placed by them on consumers' premises (whether forming the whole or part of consumers' installations or not) either belonging to them or under their control. The supply undertakers must install and maintain the lines and apparatus in a safe condition so as to prevent as far as is reasonably practicable a leakage to any adjacent metal.

Regulation 26 states that the undertakers shall not permanently connect a consumer's installation unless they are reasonably satisfied that the connection, if made, would not cause a leakage from the installation exceeding one ten-thousandth part of the maximum current to be supplied to the installation.

Regulation 27 relates to the general conditions of providing supplies to consumers in as much as the supply undertakers shall not be compelled to give a supply unless they are reasonably satisfied of the following points:

1   That all conductors and apparatus are sufficient in size and power for the purposes for which the supply of energy is to be used and are so constructed, installed and protected so as to prevent danger as far as is reasonably practicable; and that all single-pole switches are inserted in live conductors only.

2   That every distinct circuit is protected against excess energy by means of a suitable fusible cut-out or circuit breaker of adequate rupturing capacity. They must be suitably located and be of such construction as to prevent danger from overheating, arcing or the scattering of hot metal during operation. There must also be no danger when the fusible metal is renewed.

3   That every electric motor is controlled by an efficient switch or switches for starting or stopping. Switches must be so placed as to be readily accessible to and easily operated by the person in charge of the motor.

Regulation 28 forbids the supply undertakers from giving a supply at low voltage (i.e. in the regulations, a voltage less that 250 V) to any consumer from more than one pair of conductors of a three-wire, three-phase system unless the total kilowatt rating of apparatus connected to the consumers wiring exceeds 8 kW, or unless it is to avoid supply variation in excess of the limits allowed by Regulation 34 (explained on page 18). Where the supply undertakers provide a supply, they need to be satisfied that the supply terminals are arranged in separate pairs. Also, the wiring connected to the supply terminals is distinct and complies with Regulation 29, and further, in any room containing different pairs of conductors, all socket outlets are to be connected to one and the same pair of conductors.

Regulation 29 states that the supply undertakers are not compelled to give a supply at medium voltage (i.e. in these regulations a voltage exceeding 250 V but not exceeding 650 V) unless they are reasonably satisfied that all metalwork, enclosing, supporting or associated with consumers' installations, other than live conductors, is connected with earth. Also, con-

sumers' wiring must either be completely enclosed in metal which is electrically continuous and adequately protected against mechanical damage, or, alternatively, it is to be constructed, installed and protected as to prevent danger so far as is reasonably practicable.

Where a motor or separate piece of apparatus is controlled by an efficient cut-off switch, it must be readily accessible to and easily operated by the person in charge of the motor or apparatus so that the supply voltage can be cut off.

Regulation 32 is concerned with discontinuance of supply where the requirements of the regulations have been contravened.* The following provisions have effect:

1   Where the undertakers are *prima facia* satisfied that immediate action is justified as a work of emergency in the interest of public safety or in order to avoid undue interference with the efficient supply of energy to other consumers, immediate discontinuance is authorized by the regulation. The undertakers are required to give immediate notice in writing to the consumer, specifying the matter complained of.

2   In all other cases where the question of possible discontinuance arises, provision is made for due notification to the consumer of the matter complained of by the undertakers, and for the settlement of any differences that may arise between both parties in a manner provided for by Regulation 33 – notice by undertakers: procedure as to settlement of differences: appeals.

IEE Wiring Regulations for Electrical Installations

This is the current 15th Edition based upon the international rules of the IEC (International Electrotechnical Commission) and forms the basis of CENELEC (European Committee for Electrotechnical Standardization) Harmonization proposals for member states of the ECC and EFTA (European Free Trade Association).

*Regulation 31 concerns the supply to luminous signs on the outside of premises. Here, the undertakers shall not provide a supply which is transformed to a higher voltage unless the consumer gives a guarantee in writing as to the suitability and conditions of the installation.

The IEE Wiring Regulations is not a mandatory document but compliance with its Chapter 13 satisfies the Electricity Supply Regulations 1937 previously mentioned.

Basically, the regulations are designed to provide safety from fire, shock and burns and they are intended to be cited in any contract. They are not, however, intended to take the place of a detailed specification, and only established materials, equipment and materials are recognized – although new techniques and methods are not discouraged, they need only satisfy the safety requirements of the regulations.

The 15th Edition has moved a long way from earlier editions which could be practically understood by most of the electrical contracting industry workforce. Its use is now more suitable for design engineers and estimators of electrical installations, but it will still be an important document to keep on site, along with other publications and references.

A diagram showing the plan of the 15th Edition is given in Figure 1. It will be seen that it is divided into six parts, each having respective chapters and sections. Also included are seventeen appendices. The work of the IEC Technical Committee is still ongoing and some chapters are reserved for future use. The following is a summary of some of the important areas within the 15th Edition.

Part 1, Chapter 11, outlines the *scope* of the regulations and Chapter 12 its *object* and *effects*. Chapter 13 lists twenty *fundamental requirements for safety* under nine sub-headings starting with workmanship and materials and ending with inspection and testing.

Part 2 is devoted to *definitions* where a number of new terms will be found such as 'PEN conductor' and 'factory-built assembly', while other familiar terms have either been modified or have completely vanished, such as 'excess current protection' and 'earth continuity conductor'. Further reference can be made in BS 4727.

Part 3 of the 15th Edition will probably be the starting point in the design stages of an electrical installation since it is concerned with *assessment of general characteristics*. Installation parameters focus on: (i) purpose for which the installation is intended, its general structure and supply; (ii) external influence which consists of environmental conditions, type of utilization of premises and building construction; (iii) compatibility of its equipment with other electrical equipment and other services including impairment of the supply through, for example, starting currents or even high frequency oscillations; and also maintainability of the installation during its intended life. Having gathered this information the installation design moves forward to estimating the maximum demand required. Some guidance in this area is given in Appendix 4 but reference should be made to the notes of Table 4B. With the necessary data and diversity allowances (if any), information is then required about the arrangements of live conductors and type of earthing as well as nature of supply and installation circuit arrangements. Here, consultation is required with the supply undertaking and consideration given to supplies for safety services and standby purposes as well as circuit planning. The installation designer must not lose sight of the maintenance requirements where the quality of the equipment and materials used must be carefully selected if they are to ensure reliability of operation for their intended life. Chapter 35 is a new chapter concerned with supplies for safety services (see also 'Standby supplies', page 70).

Part 4 deals with *protection for safety* and covers protection against electric shock, thermal effects and overcurrent, also isolation and switching and application of protective measures for safety. This is an important area for the designer since he is able to identify and specify the protection methods he intends to use. Requirements 471–2 to 471–5 concern protection against both direct and indirect contact which are often regarded as the two major lines of defence against electric shock. They are further treated in requirements 471–6 to 471–10 and 471–11 to 471–21, respectively. Sections 472, 474 and 475 are reserved for future use.

Part 5 is concerned with *selection and erection of equipment* and covers a number of chapters more appropriate to the electrical installer. Briefly, there are: common rules; cables, conductors and wiring materials; switchgear; earthing arrangements and protective conductors; and other equipment. Several appendices are required to be used in this part of the regulations. Chapter 56 deals with supplies for safety services in detail.

Part 6 concentrates on *inspection and testing* with

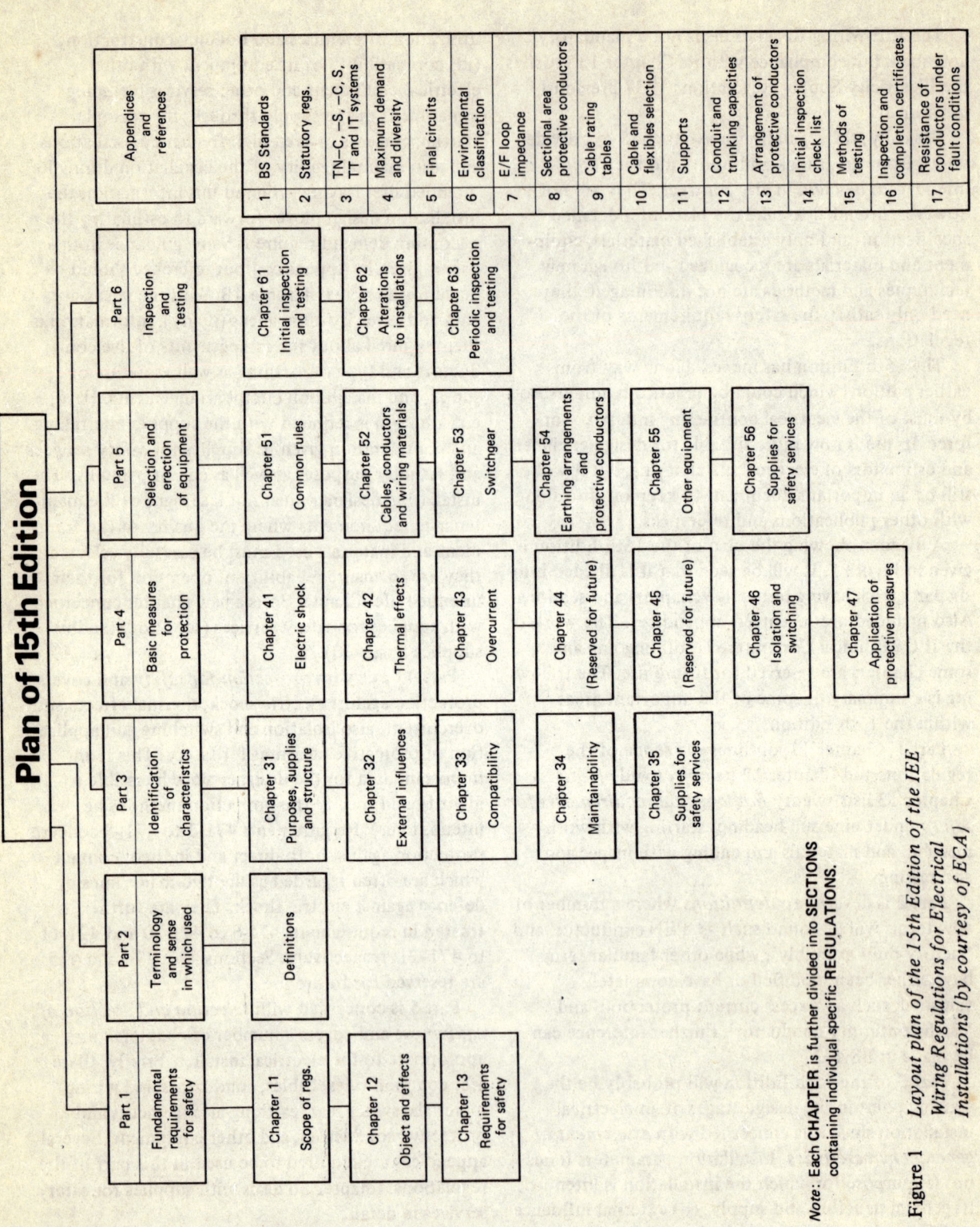

Plan of 15th Edition

Part 1 Fundamental requirements for safety	Part 2 Terminology and sense in which used	Part 3 Identification of characteristics	Part 4 Basic measures for protection	Part 5 Selection and erection of equipment	Part 6 Inspection and testing	Appendices and references

Chapter 11 Scope of regs.	Definitions	Chapter 31 Purposes, supplies and structure	Chapter 41 Electric shock	Chapter 51 Common rules	Chapter 61 Initial inspection and testing	1 BS Standards
Chapter 12 Object and effects		Chapter 32 External influences	Chapter 42 Thermal effects	Chapter 52 Cables, conductors and wiring materials	Chapter 62 Alterations to installations	2 Statutory regs.
Chapter 13 Requirements for safety		Chapter 33 Compatibility	Chapter 43 Overcurrent	Chapter 53 Switchgear	Chapter 63 Periodic inspection and testing	3 TN–C, –S, –C, S, TT and IT systems
		Chapter 34 Maintainability	Chapter 44 (Reserved for future)	Chapter 54 Earthing arrangements and protective conductors		4 Maximum demand and diversity
		Chapter 35 Supplies for safety services	Chapter 45 (Reserved for future)	Chapter 55 Other equipment		5 Final circuits
			Chapter 46 Isolation and switching	Chapter 56 Supplies for safety services		6 Environmental classification
			Chapter 47 Application of protective measures			7 E/F loop impedance
						8 Sectional area protective conductors
						9 Cable rating tables
						10 Cable and flexibles selection
						11 Supports
						12 Conduit and trunking capacities
						13 Arrangement of protective conductors
						14 Initial inspection check list
						15 Methods of testing
						16 Inspection and completion certificates
						17 Resistance of conductors under fault conditions

Note: Each **CHAPTER** is further divided into **SECTIONS** containing individual specific **REGULATIONS**.

Figure 1   *Layout plan of the 15th Edition of the IEE Wiring Regulations for Electrical Installations (by courtesy of ECA)*

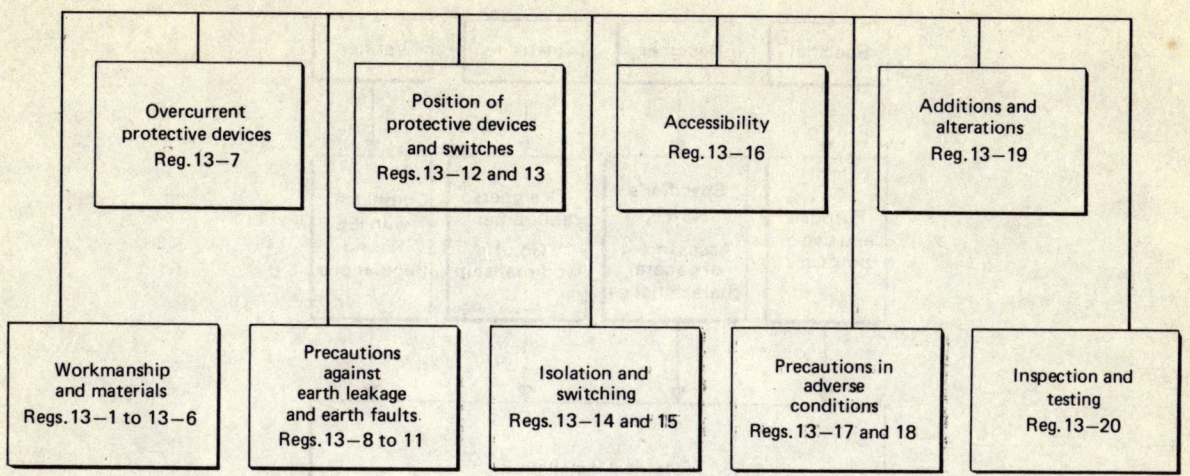

Figure 2   *Block diagram representing Chapter 13 of the IEE Wiring Regulations – Fundamental requirements for safety*

inspection being carried out to ascertain that all equipment complies with relevant British Standards (or equivalent foreign standards) and that it is correctly selected in accordance with the Regulations and is not visibly damaged so as to impair its safety. Appendix 14 provides guidance on the items that require checking, such as the presence of fire barriers and presence of danger notices. The purpose of testing is to ensure that the requirements of the Regulations have been met and calls for tests ranging from continuity of ring final circuit conductors to operation of earth leakage circuit breakers. Notes on the standard methods of testing are given in Appendix 15. Part 6 also covers certification with reference made to Appendix 16 and there are two further chapters concerning alterations to installations and periodic inspection and testing, again reference being made to Appendix 16.

Figure 2 illustrates the fundamental requirements for safety while Figure 3 shows the responsible parties associated with an electrical installation project.

British Standards

These are often described as *technical agreements* published by the British Standards Institution (BSI)

which was set up in 1901 with the intention of creating national standards for the engineering industry. Except in a few cases where the government has produced regulation documents such as The Electric Blanket (Safety) Regulations 1971 and The Heating Appliances (Fireguards) Regulations 1973, making it mandatory for manufacturers to use certain British Standards, the vast majority are in fact of a voluntary nature. There seem to be British Standards for practically everything made today, each having its own code number and incorporating sufficient details about a particular product to meet certain specifications.

In dealing with requirements for safety, BSI have created two important marks of recognition for certain products, namely, a certification trade mark better known as the BSI kite-mark and a BSI safety-mark. Both of these are shown in Figure 4. The safety-mark stems from an EEC Low Voltage Directive which paved the way for the setting up of the Electrical Equipment (Safety) Regulations 1975. It is a guarantee of a product's electrical, mechanical and thermal safety, although not necessarily performance. The BSI kite-mark is an assurance that a product has been produced under a system of supervision, control and testing and is only used by those manufacturers

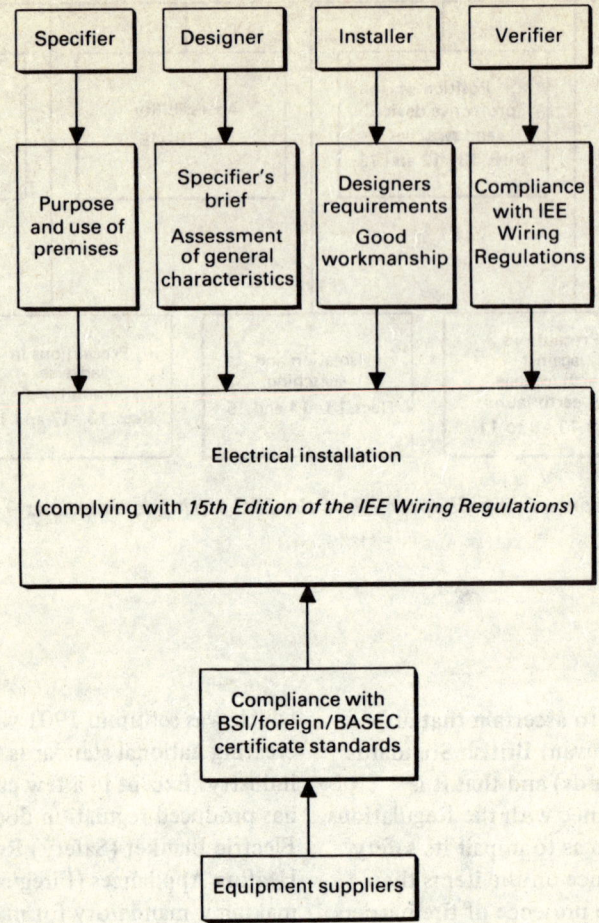

Figure 3   *Responsible parties in the development process of an electrical installation*

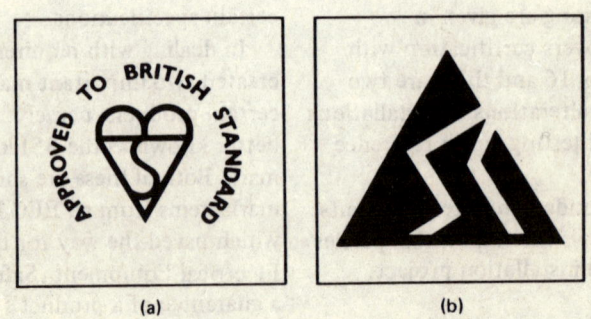

Figure 4   *BSI markings*
         (a) Kite-mark
         (b) Safety-mark

granted a licence under the scheme. It may not cover safety unless the appropriate British Standard includes the necessary requirements.

In Appendix 1 of the IEE Wiring Regulations, mention is made to some sixty-eight British Standards together with eight British Standards Codes of Practice. Some of these such as BS 5839: 'Fire detection and alarm systems in buildings' will receive a mention throughout the pages of this book – this also applies to numerous IEE Wiring Regulations requirements.

Additional requirements

Chapter 2 gives further mention to statutory requirements when dealing with supply variation, substations and earthing and more requirements for safety will be found in Chapter 3, particularly in connection with electrical apparatus in hazardous areas. Probably one of the earliest requirements for flammable atmospheres came with the Electricity (Factories Act) Special Regulations of 1908–1944 where it is stated in Regulation 27 that:

'All conductors and apparatus exposed to . . . flammable surroundings or explosive atmosphere . . . shall be so constructed and protected, and such special precautions shall be taken as may be necessary adequately to prevent danger in view of such exposure or use.'

Unfortunately, the regulation does not give any mention of standards or certification, or even the approval of special apparatus for use in such explosive atmospheres, but despite this, practice of using tested and certified equipment has been carried out for some years. The body which has become recognized for this work is known as BASEEFA – the British Approvals Service for Electrical Equipment in Flammable Atmospheres. It was formed in 1967 and is now administered by the Health and Safety Commission (HSC). BASEEFA is the British national testing and certification authority and is one of the test houses for European Directives concerned with flammable atmospheres for the surface industries. In 1978, BASEEFA became self-reliant for all mainstream testing of electrical equipment. It was arranged in three sections, namely: (i) heavy current and reliability concepts; (ii) light current and intrinsic

safety; and (iii) flameproof and pressurization. Any first-time manufacturer intending to design electrical equipment for use in flammable atmospheres will need to decide which method of protection is best suited for his product since there are numerous methods available which Chapter 2 attempts to uncover. In practice, however, certified equipment by BASEEFA will be able to bear both a certificate number and mark of approval in the shape of a crown with the inscription 'Ex' signifying *explosion protection* (see Figure 5).

Chapter 3 provides more safety requirements, particularly in petrol filling stations, garages, farms, fire alarm systems and standby supplies. In this last case, for example, emergency lighting is governed by several national and local regulations and the electrical contractor has to be aware of these. For instance, there is the 1976 Fire Certificates (Special Premises) Regulations concerned primarily with hotels and boarding houses and giving widespread publicity on the subject of safety inside these types of premises. Other buildings are covered by the Offices, Shops and Railway Premises Act of 1963, cinemas and theatres by Acts of 1955 and 1968, respectively, while other places of entertainment are covered in an Act of 1967. When planning emergency lighting, assistance should always be sought from the local fire brigade.

From the point of view of the Health and Safety at Work etc. Act, under Section 6 of this Act, existing regulations such as the Electricity (Factories Act) Special Regulations and Factories Act, are retained, and at present are being revised. It should be pointed out, therefore, that the electrical contractor, where he has control of the installation, is in fact seen as the

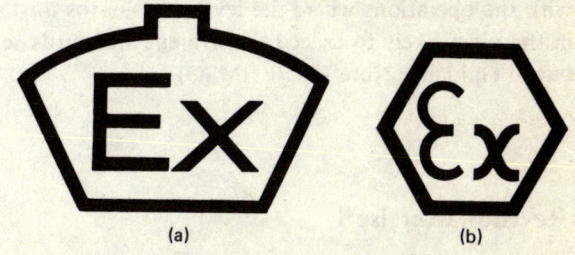

Figure 5 *Explosive equipment markings*
(a) BASEEFA mark
(b) EEC mark

occupier which imposes on him duties under the Health and Safety Act to provide not only safe systems of work, but also provision of information, instruction, training and supervision. Similarly, there is a duty imposed on his workforce to conduct their work in accordance with the Act and electrical requirements. The main duties required by the Health and Safety Act for both employer and employee are mentioned in Volume 2 of this series. It is important to have trained staff to deal with matters concerning safety and site operatives should be acquainted with company safety procedures such as the wearing of protective clothing for personal safety, and the safe use of power tools, such as cartridge fixing tools, which call for constant attention when loaded.

A code has been prepared for practical guidance in respect of the Asbestos Regulations 1969. This code, namely, 'Work with asbestos insulation and asbestos coating' (1981) deals with all aspects of working with asbestos insulation and sprayed coatings and it includes matters such as informing workers of the health risks associated with asbestos and ensuring that they are trained to use personal protective equipment, as well as being familiar with techniques used to prevent asbestos fibres escaping from working areas. The code also advises on the disposal of asbestos waste, which should consider the provision of a sufficient number of suitable waste receptacles, transfer of waste to the disposal point and action to be taken in the event of spillage. Under the 1969 legislation (Reg. 20) and Section 2(1) and 2(2) of the Health and Safety Act, employees under the age of eighteen should not be employed in any operation involving the treatment and removal of asbestos insulation, the bagging of waste products removed in stripping operations or any other work associated with the operations where the level of asbestos dust in the air is likely to exceed the hygiene standards set out in Guidance Note EH10(HMSO).

Revision exercise 1

1   Describe the requirements of the Electricity Supply Regulations in respect of:
    (a) entry of underground cables
    (b) leakage of current to earth

(c) equipment placed on the premises by the undertaking
(d) protection of consumers' installations against excess energy

CGLI/C/82

2   With reference being made to the Electricity Supply Regulations 1937:
    (a) determine the minimum permissible insulation resistance of a domestic consumer's wiring if his maximum supply current is 100 A
    (b) state the lettering or colouration used for identifying the conductors of a main service cable in a TN–C–S system
    (c) determine the supply voltage and frequency variation limits on a consumer's premises supplied at 415 V/50 Hz

3   Explain the *reasons* for the IEE Wiring Regulations regarding the following:
    (a) limitations of voltage drop in consumers' final circuits
    (b) knowing the value of prospective short-circuit current at the origin of an installation
    (c) switching off for mechanical maintenance

4   (a) Make a block diagram from the IEE Wiring Regulations which illustrates Section 471 – Protection against electric shock.
    (b) Distinguish between protection against direct contact and protection against indirect contact, giving *two* examples in each case.

5   With reference to the IEE Wiring Regulations, explain the reasons behind the following regulations:
    (i)   471–34
    (ii)   473–2
    (iii)   530–2
    (iv)   531–4
    (v)   554–39

6   Write a brief account of any *one* of the following:
    (i)   BS 2754
    (ii)   BS 3052
    (iii)   BS 3456
    (iv)   BS 3535
    (v)   BS 5266

7   (a) A complaint is received that an electric
        washing machine in a laundrette is giving
        shocks to users. It is found that a current
        operated earth leakage circuit-breaker
        (residual current device) is installed in each
        washing machine circuit to protect users.
        Detail *three* tests which may be carried out
        to establish the cause of this defect.
    (b) State *two* advantages and *one* disadvantage
        of a current operated earth leakage circuit-
        breaker (residual current device) in compari-
        son with a fault voltage operated device for
        premises as in (a) above.
                                          CGLI/C/82

8   (a) Under what circumstances may a supply
        undertaking refuse to:
        (i) provide a supply of electricity to a new
             installation
        (ii) continue a supply of electricity to an
             existing installation?
    (b) What action must be taken by the supply
        undertaking in the two cases described in (a)?
    (c) What action may the consumer take if the
        refusal to continue a supply of electricity to
        an existing installation is considered to be
        unjustified?
                                          CGLI/C/84

9   (a) Explain why it is essential to request certain
        details from the supply undertaking before
        planning an installation.
    (b) State the requirements of the Electricity
        Supply Regulations with regard to
        information which the supply undertaking
        must furnish.
    (c) What additional details are essential to allow
        compliance with the IEE Regulations?
                                          CGLI/C/84

10 The chargehand of a team of electricians is
    informed that a member of the team has fallen
    whilst ascending a ladder and has been taken to
    hospital. State the main checks to be made before
    writing a report to the employer about the
    following factors:
    (a) witnesses
    (b) the ladder
    (c) the fitness of the victim before the accident
                                          CGLI/C/84

chapter two

Supply and distribution of electricity

After reading this chapter you will be able to:

1 State the requirements for a supply authority with regard to consumers' supply voltage and frequency variations.

2 Describe, with the aid of a diagram, voltage regulation.

3 State a number of requirements concerning sub-stations.

4 Draw a labelled diagram of a substation layout showing the siting of electrical equipment.

5 Perform simple short-circuit calculations.

6 Describe the advantages of XLPE as a cable insulant and know types of cable using this material.

7 Perform calculations to ascertain cable size and voltage drop.

8 Know requirements for earthing a consumer's premises and draw various types of distribution systems.

9 Describe various factors associated with the design and installation of large consumer premises.

10 Perform numerous revision questions.

Electricity supply

Whereas the *generation* and *transmission* of electricity in England and Wales is the prime responsibility of the Central Electricity Generating Board (CEGB), *distribution* of electricity is the prime responsibility of twelve Area Electricity Boards who purchase bulk supplies from the CEGB.

Generation of electricity in the majority of the CEGB's power stations is in the order of 25 kV and much of their transmission lines operate at 275 kV and 400 kV. This high voltage primary transmission is known as the 'Supergrid' as shown in Figure 6. It can be seen that much of the 275 kV concentrates around distinct *load centres* which are areas where bulk supplies are required, such as London and Birmingham. The 400 kV lines are used to link up new power stations and tapping points which are dispersed around some 5000 route kilometres.

To supervise the operation of power stations feeding into the Grid, five regional organizations, known as *control centres* are set up, and they are in direct communication with the power stations and National Control Centre in London. The control centres are not only responsible for generation and transmission at regional level, which includes operation and maintenance, but also the secondary transmission connections at 132 kV and less, supplied to Area Boards and large industrial users.

Supply authorities are permitted, under the Electricity Supply Regulations 1937 (Reg. 34b), to allow consumers' voltage to vary by not more than ± 6 per cent and supply frequency by not more than ± 1 per cent. Thus, for a domestic consumer supplied at 240 V, 50 Hz, the voltage variation is between 225.6 V and 254.4 V while frequency is allowed to vary between 49.5 Hz and 50.5 Hz. These limitations

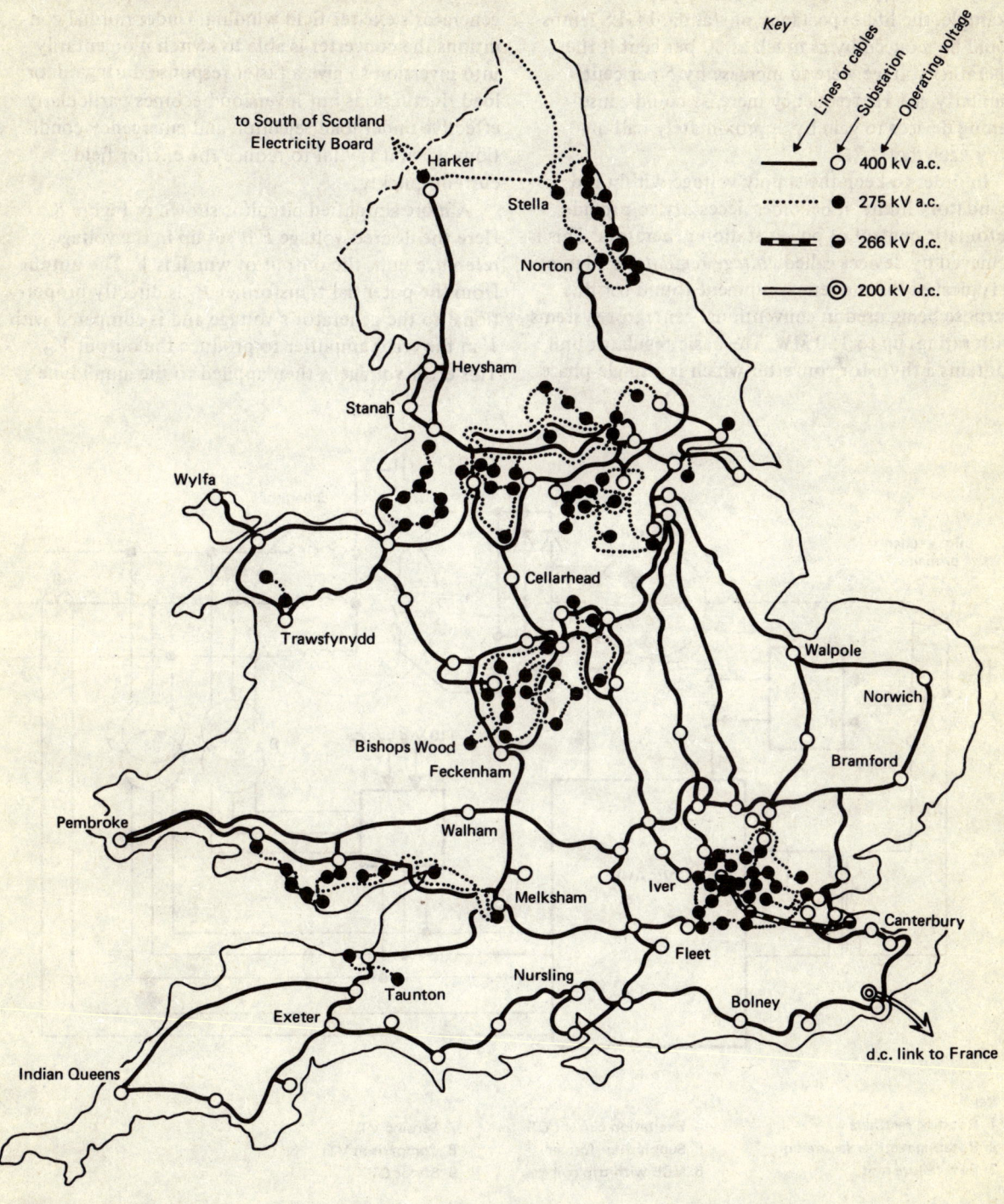

Key

Lines or cables
Substation
Operating voltage

○ 400 kV a.c.

● 275 kV a.c.

⊖ 266 kV d.c.

◎ 200 kV d.c.

to South of Scotland
Electricity Board

Harker

Stella

Norton

Heysham

Stanah

Wylfa

Trawsfynydd

Cellarhead

Walpole

Norwich

Bishops Wood

Feckenham

Bramford

Pembroke

Walham

Melksham

Iver

Canterbury

Taunton

Nursling

Fleet

Exeter

Bolney

Indian Queens

d.c. link to France

Figure 6   *The supergrid transmission system in England and Wales*

are very important because any marked variation is likely to damage or upset electrical equipment. For example, the life expectancy of standard GLS lamps could be reduced by as much as 50 per cent if the operating voltage were to increase by 5 per cent. Similarly, a 1 Hz frequency increase could cause timing devices to gain by approximately half-an-hour each day.

In order to keep the supply voltage within the mandatory limits, it becomes necessary to provide automatic control of power station generators. This is achieved by devices called *voltage regulators*. Figure 7 is typical of the modern equipment found for this purpose being used in conventional generator systems with ratings up to 150 MW. The basic regulator unit contains a thyristor converter which is a single-phase

full-wave unit capable of supplying the necessary changes in output current demanded by the generator's exciter field winding. Under normal conditions the converter is able to switch momentarily into inversion to give a faster response during minor load fluctuations but inversion becomes particularly effective under load rejection and emergency conditions when it is vital to reduce the exciter field current quickly.

A more simplified circuit is shown in Figure 8. Here the desired voltage E is set up in the voltage reference unit, the output of which is V. The output from the potential transformer E_t is directly proportional to the generator's voltage and is compared with V in the error amplifier to produce the output V_o. This error voltage is then applied to the amplidyne's

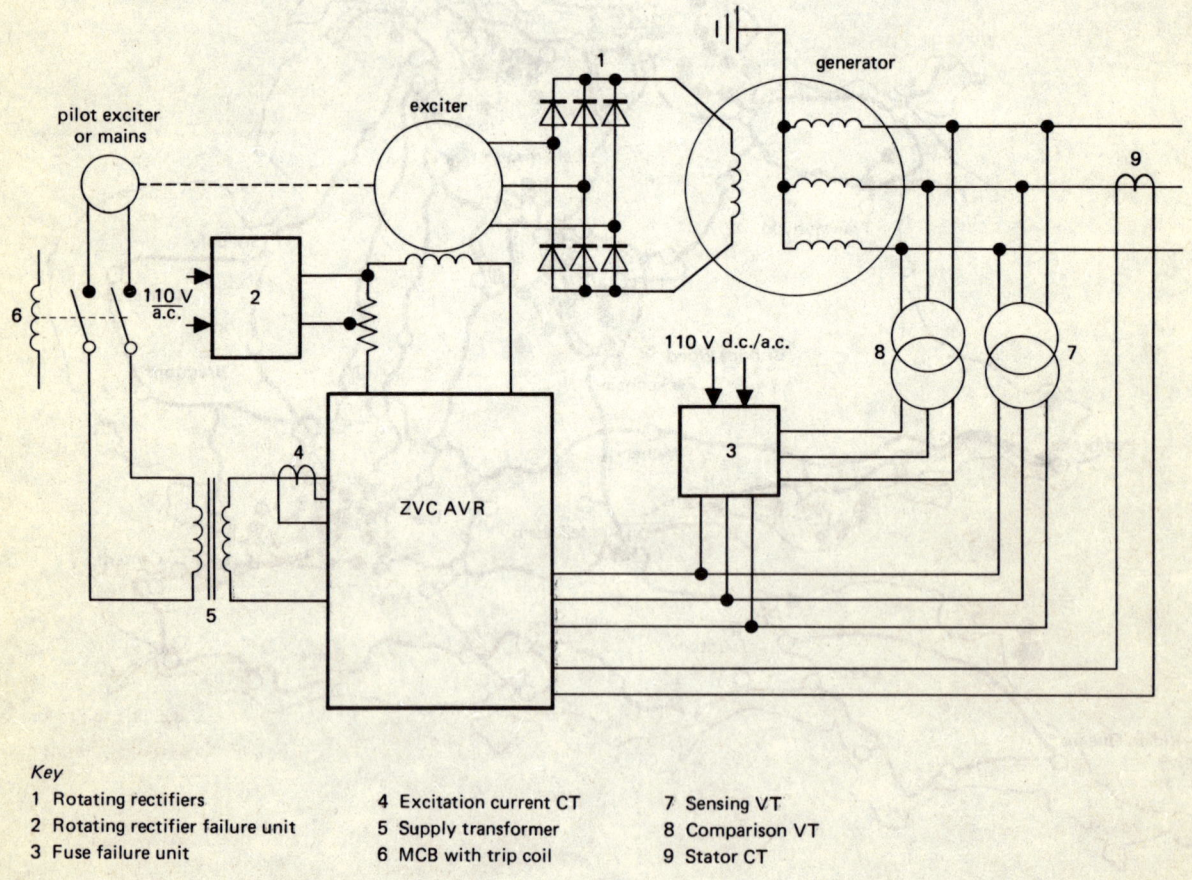

Key

1 Rotating rectifiers	4 Excitation current CT	7 Sensing VT
2 Rotating rectifier failure unit	5 Supply transformer	8 Comparison VT
3 Fuse failure unit	6 MCB with trip coil	9 Stator CT

Figure 7   *ZVC series excitation control system (by courtesy of GEC Industrial Controls Ltd)*

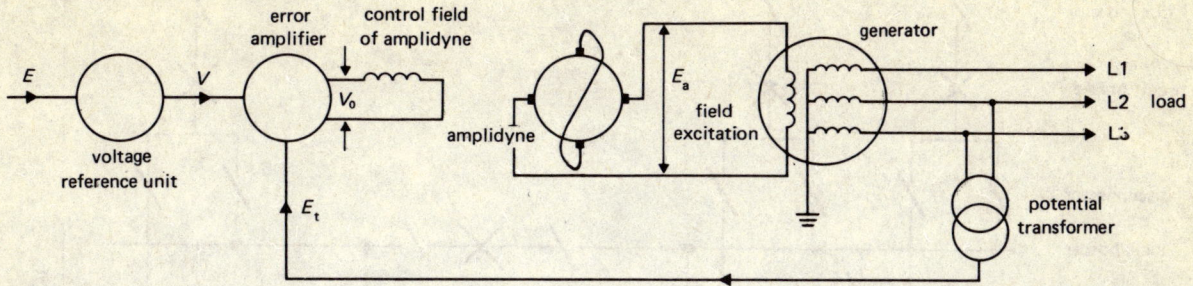

Figure 8   *Voltage regulator*

control field which results in an amplidyne output voltage E_a. This voltage then acts as the control signal to the generator's field. Thus, once the system has been set to its steady-state condition corresponding to the required generator voltage, then any variation in that voltage will result in a feedback signal which will act in opposition to these variations.

Substations

Consumers' electrical distribution starts at the intake position on their premises. Before this arrangement is reached the system's voltage would have probably been reduced by one of the many CEGB substations. These substations, often complex, contain apparatus for transforming or converting energy to or from a voltage above low voltage and they may include other apparatus for switching and controlling the energy as part of their function. They are often sited near the *centre of gravity* of the potential load. In air-insulated outdoor open type substations handling 400 kV, large clearances are required around the electrical busbars to avoid the likelihood of flashover. Regulation 9 of the 1937 Electricity Supply Regulations requires such stations to be efficiently protected by fencing not less than eight feet (2.44 m) high or other means so as to prevent access to the electric lines and apparatus therein by any unauthorized person. In large cities the substations are mostly enclosed in large buildings. A typical layout of the electrical equipment for a 400/132 kV substation is shown in Figure 9. It comprises incoming and outgoing circuits connected through switchgear to common busbars which are often duplicated to allow the system to operate in the event of faults occurring or maintenance work being

carried out. For this reason it will be seen that various sections of equipment can be isolated, then made safe by earthing, the function of the switchgear being to disconnect the circuit elements as and when required. Over the years, as generating capacity and power flow have increased, switchgear has been required to clear higher levels of fault. The latest types of oil and gas-blast circuit breakers are now able to handle fault levels up to 35000 MVA.

Figure 10 is a typical substation/switching station layout catering for 33/11 kV. It shows the switchgear protection on either side of a single main busbar arrangement. Included on the diagram are current transformers to measure the power flow and detect overload and fault conditions through protective equipment (not shown) which operate the circuit breakers, also voltage transformers used for measuring the voltage conditions within the system. Like the previous illustration (Figure 9) it will be noticed that there is adequate means available for isolating parts of the system for emergency and maintenance repair work.

There are two important regulation requirements concerning substations and these are given by the Electricity (Factories Act) Special Regulations 1908 and 1944. They are:

Regulation 30 which states:

'Every substation shall be substantially constructed and shall be so arranged that no person other than an authorized person can obtain access thereto otherwise than by the proper entrance, or can interfere with the apparatus or conductors therein from outside: and shall be provided with efficient means of ventilation and kept dry.'

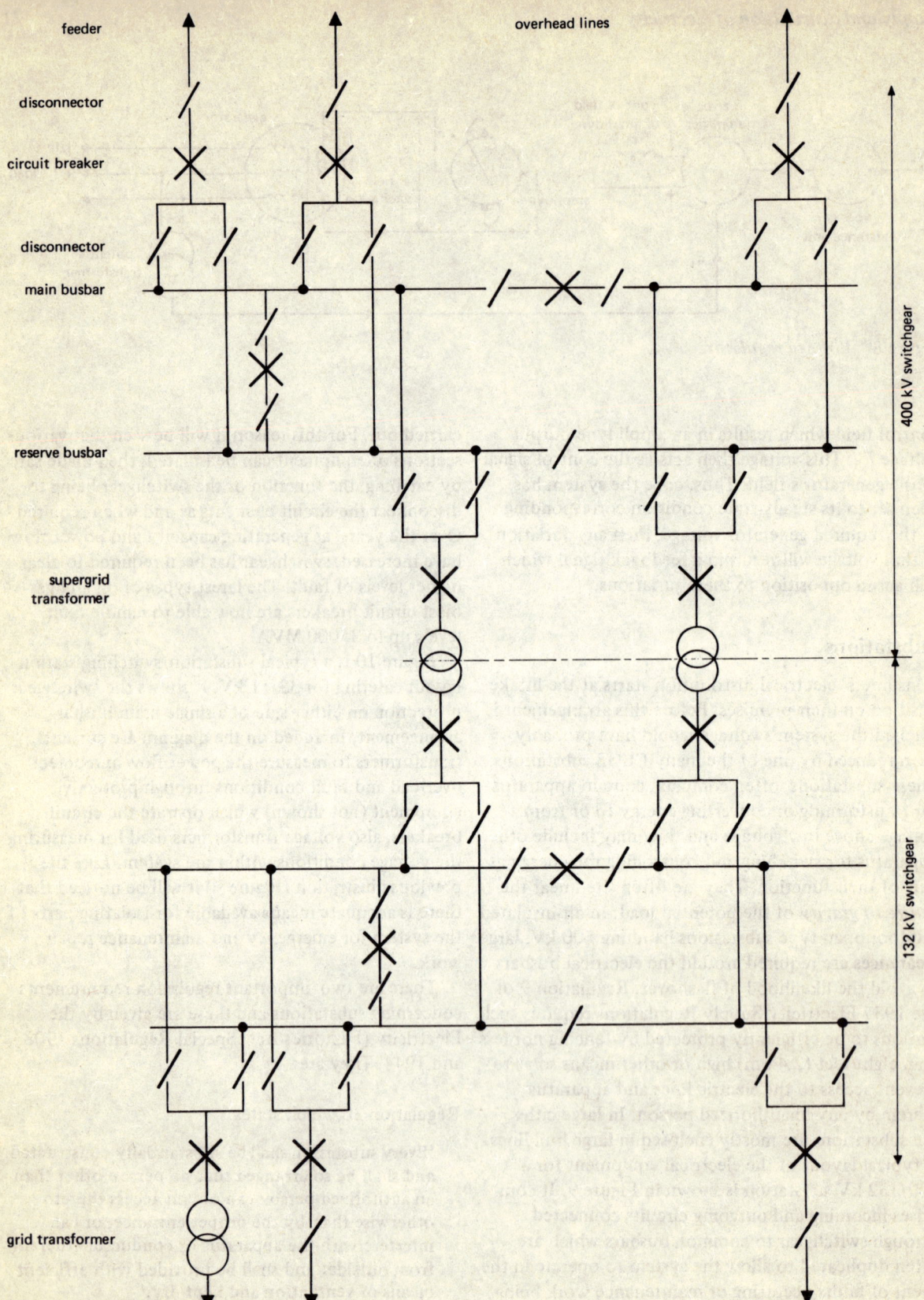

feeder

overhead lines

disconnector

circuit breaker

disconnector

main busbar

reserve busbar

supergrid transformer

grid transformer

400 kV switchgear

132 kV switchgear

Figure 9   *400 kV/132 kV substation*

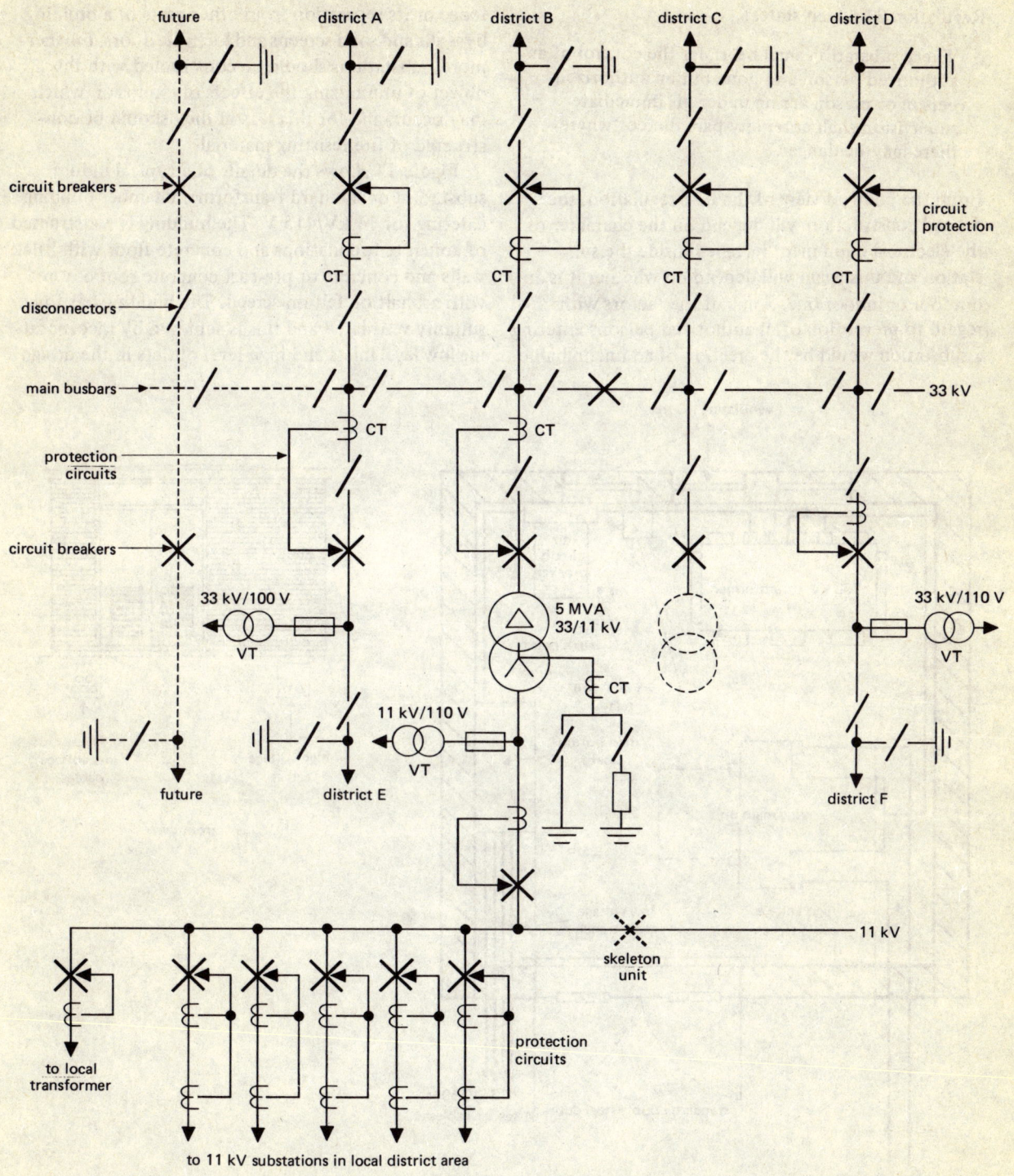

Figure 10   *33 kV/11 kV substation layout*

Regulation 31 which states:

> 'Every substation shall be under the control of an authorized person and none but an authorized person or person acting under his immediate supervision shall enter any part thereof where there may be danger.'

From the point of view of the first regulation, the general construction will depend on the character of the electrical equipment installed inside the substation and this again will depend on whether it is an outdoor or indoor one. Some of the factors with regard to prevention of unauthorized persons entering a substation would be the erection of an unclimbable

fence or its separation from other parts of a building by walls and solid screens and lockable doors. Furthermore, substations should be constructed with the object of minimizing the effects of explosion which may occur, and for this reason they should be constructed of fire resisting material.

Figure 11 shows the details of a typical indoor substation or standard transformer chamber building catering for 11 kV/415 V. The building is constructed of concrete foundations and concrete floor with brick walls and concrete or pre-cast concrete roof covered with asphalt on felt on screed. The building has to be suitably ventilated and this is achieved by incorporating low level inlets and high level outlets in the design.

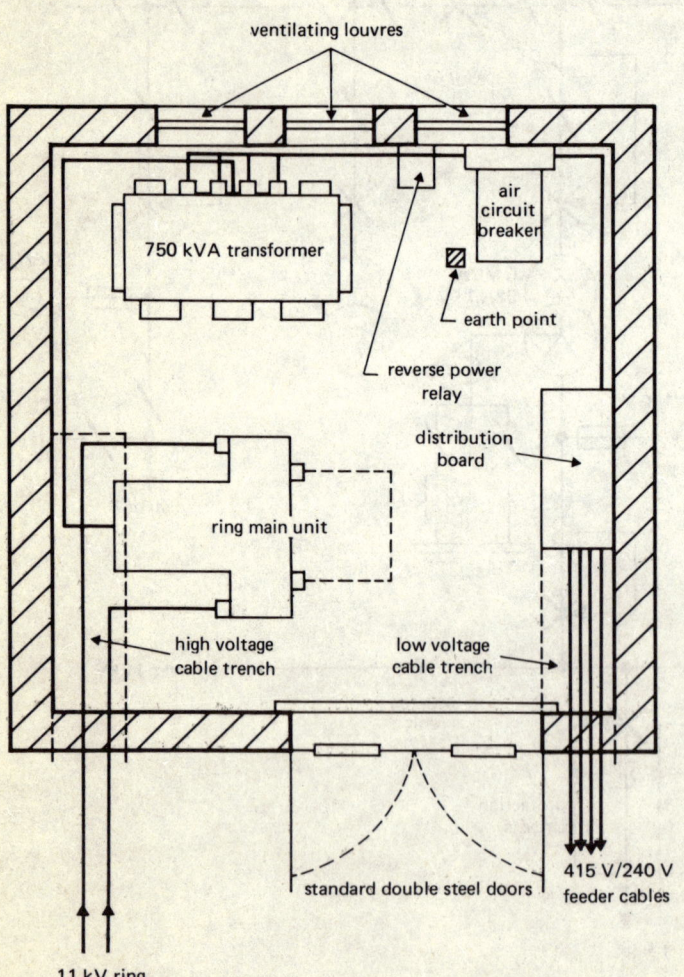
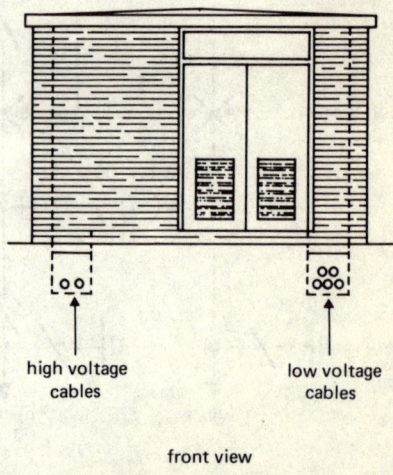

Figure 11   *Typical standard transformer chamber building*

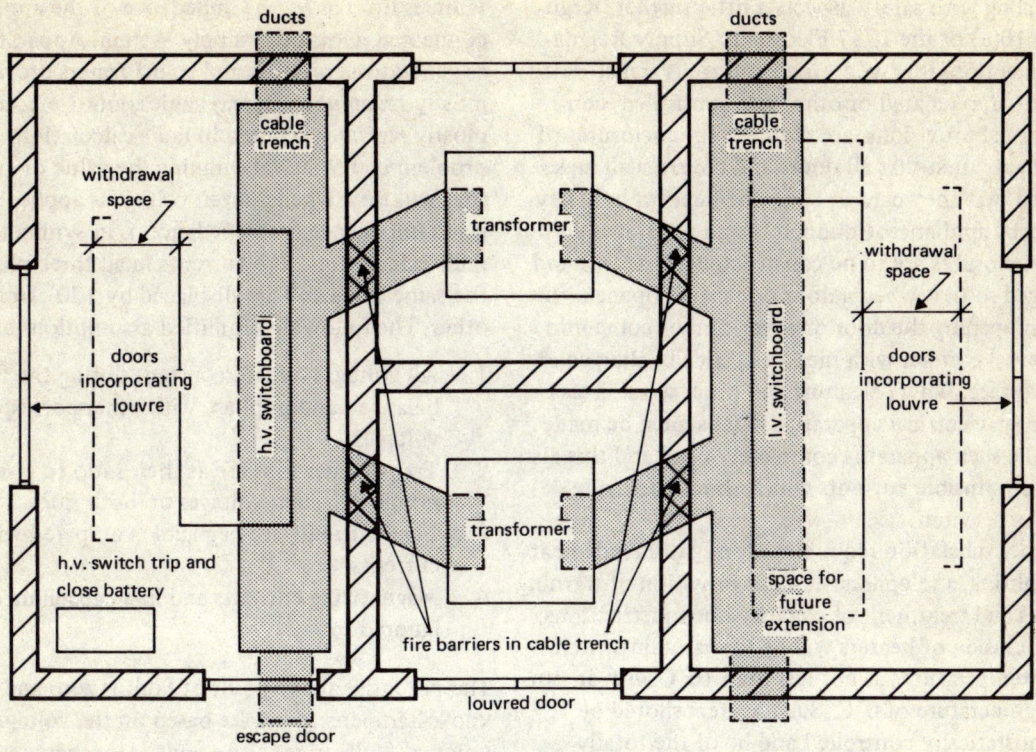

Figure 12 *High voltage substation comprising two 500 kVA delta-star transformers for low voltage distribution*

Also, the high voltage switchgear needs to be separated from the low voltage switchgear. In larger capacity substations, transformers are often kept separated from other apparatus by erecting brick dividing walls. Doors, where they are fitted, should be self-closing and fire proof. The main doors are of steel construction and open outwards to enable personnel to escape in the event of a fire or explosion. If doors are constructed of wood, they should be treated with fire-resistant paint as should all other woodwork inside the substation. In basement type substations, consideration needs to be given to precautions taken against risk of fire as laid down by the 1937 Electricity Supply Regulations, Regulation 16. The requirement concerns apparatus such as oil-immersed transformers and switches that use in excess of 2000 gallons (9092 litres) of oil in any one tank, receptacle or chamber. Provisions have to be made for the draining away or removal of any oil which may

leak or escape from the apparatus. One method is to place tanks etc. above a pit containing ballast or graded chippings. In basement, and other types of substation that are indoors, it may be necessary to provide automatic fire extinguishing methods, such as the 'total flooding system' using carbon dioxide gas as the extinguishing agent.

The other Factory Act regulation (Reg. 31) concerns the proper control of a substation such that no person other than an authorized person or person acting under his immediate supervision shall be allowed to enter any part of the substation where there may be danger. In cases where the supply authorities have substations on consumers' premises, the occupier of a building may have access for switching and other purposes, and here the supply authorities need to be satisfied that the consumer or his representative are competent to carry out work as well as avoid any danger that may exist.

Dealing with safety aspects a little further, Regulation 10(a) of the 1937 Electricity Supply Regulations mentions that where high voltage is transformed, converted, regulated or otherwise controlled, constructional provisions are required for enclosures of switchgear such that all doors and covers shall be so secured that they cannot be opened except by a key or special appliance. Conductors and apparatus within such enclosures are to be constructed, protected and arranged so that when a door or cover is opened, the person opening the door or cover cannot come into accidental contact with metal electrically charged at high voltage. Where cleaning and other work is required on electrical apparatus, means must be made to make such apparatus completely dead and this also applies to fusible cut-outs which should be made dead by a switch.

Other substation requirements might include heating, lighting, a telephone and the provision of warning notices and treatment of electric shock instructions. The inclusion of heaters will be to maintain switchroom temperatures at not less than 10°C with an outside temperature of 0°C. Such heaters should be thermostatically controlled and be of the totally enclosed, low temperature type. Adequate lighting should be provided, particularly in the high voltage switchroom area and around the transformer chamber. The recommended illuminance level is 150 lux. Finally, there should be an ample supply of 13 A socket outlets distributed around the building. Figure 12 shows the essential switchgear and layout for a high voltage substation comprising two 500 kVA delta-star transformers.

Short-circuit calculations

On the supply side of consumers' premises, it is often necessary to carry out short-circuit calculations in order to make proper selection of protective apparatus. During a short circuit, protective apparatus has to control relatively large amounts of electrical energy, and apart from the thermal effects, which often cause insulation breakdown, there are mechanical and electromagnetic effects which place tremendous strain on fixings and supports.

It is often the function of circuit breakers connected in the transmission network which are left to clear short-circuit faults. The magnitude of these faults is limited by the impedance of the apparatus connected within the supply system. Apparatus such as generators, reactors and transformers provide mostly *reactance* whereas cables, joints etc. provide mostly *resistance*. Often in fault calculations the problem is that of determining the value of current that will flow when a given voltage is applied to a given impedance (or impedances). In symmetrical-fault calculations, the currents in all three phases have the same value and are displaced by 120° from each other. The following simplified assumptions are made:

1   All voltages remain constant during the fault, balanced and in phase with other corresponding voltages.
2   Transformer taps are at their ratio to give nominal system voltages on both sides.
3   Load currents are negligible compared with fault currents.
4   Magnetizing currents and line capacitances are ignored.

The potential at the point of fault is zero and the kilovoltamperes given are based on the voltage at the point of fault prior to the fault. Also, percentage values are used in calculations, these being determined by the rated voltamperes of switchgear/apparatus. If all apparatus is referred to a common base and a common voltage assumed, the percentage values can be added and the system reduced to a source of voltage in series with a percentage impedance. It will be seen in Figure 13 that the voltage rises from zero at fault to source voltage in a manner determined by the percentage impedance involved. The following examples show how these calculations are carried out.

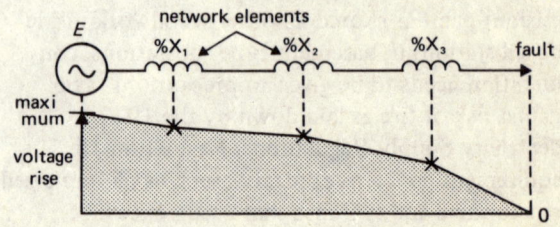

Figure 13   *Graph of voltage across network elements*

Example 1

In Figure 14, a symmetrical three-phase fault occurs*
on one of the feeders (F). Find the fault MVA and
the fault current.

   As previously mentioned, all elements must be
brought to a common base rating and this we will
choose as 150 MVA. The new value of reactance (X)
in each case will be found from:

$$\text{New } X = \frac{\text{Rated } X \times \text{Base VA}}{\text{Rated VA}}$$

Thus, for G1 (30 MVA, 15 per cent X), the new value
of X is uprated and in per cent form is

$$\%X = \frac{15 \times 150}{30} = 75\%$$

For the other elements:-

G2 $\%X = \dfrac{20 \times 150}{50} = 60\%$

G3 $\%X = \dfrac{18 \times 150}{30} = 90\%$

R $\%X = \dfrac{5 \times 150}{20} = 37.5\%$

T1 $\%X = \dfrac{8 \times 150}{25} = 48\%$

T2 $\%X = \dfrac{7 \times 150}{20} = 52.5\%$

T3 $\%X = \dfrac{6 \times 150}{15} = 60\%$

T4 $\%X = \dfrac{5.5 \times 150}{5} = 165\%$

For the line it is assumed to have 2 ohms/phase (its
resistance will not be considered in this calculation)
and its new per cent X value is found by

Line $\%X = \dfrac{\text{VA base} \times 100 \times \text{Rated } X}{\text{Line volts}^2}$

$$= \frac{150 \times 100 \times 2 \times 10^6}{33^2 \times 10^6} = 27.5\%$$

Grid $\%X = \dfrac{\text{VA base} \times 100}{\text{Short-circuit rating}}$

$$= \frac{150 \times 100}{1000} = 15\%$$

*A symmetrical three-phase fault is one that is a direct short
across all live conductors.

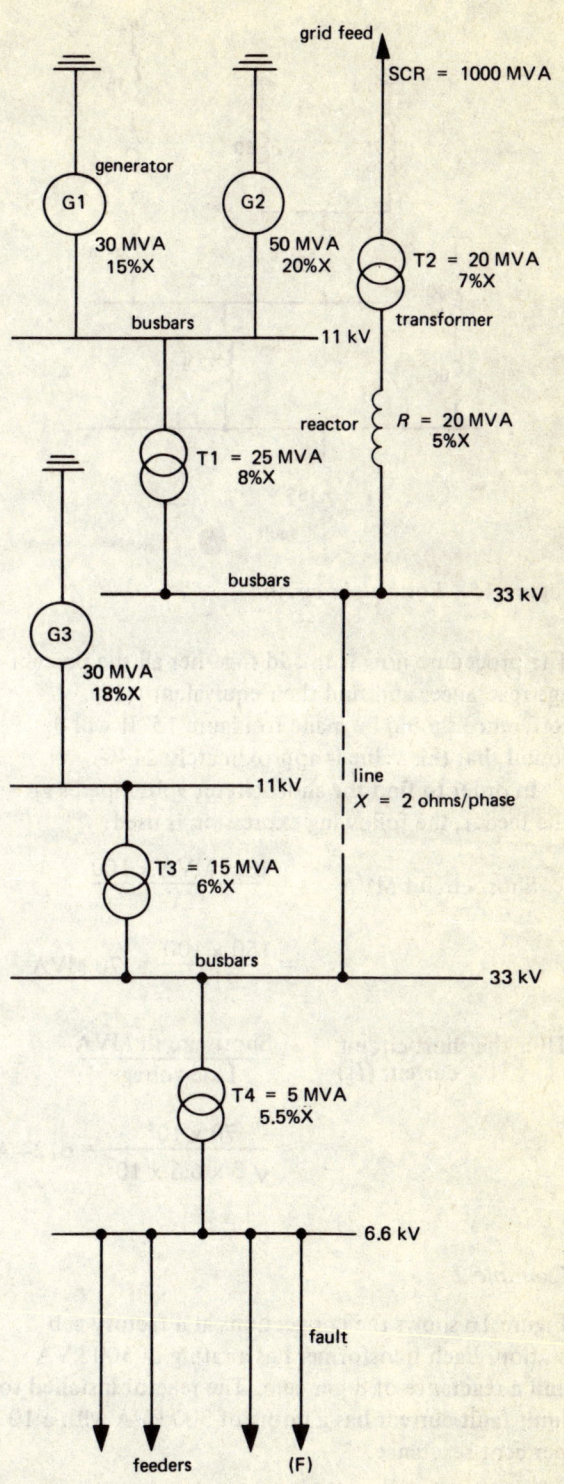

Figure 14

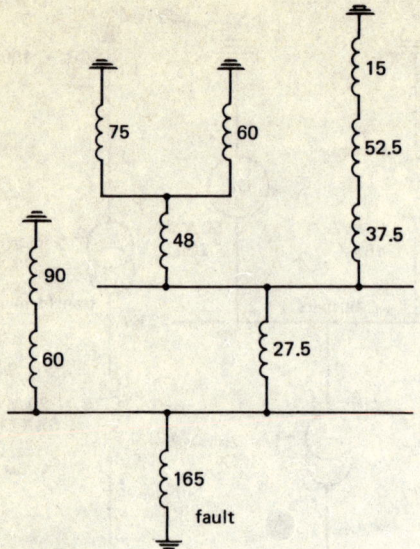

Figure 15   *Equivalent circuit*

The procedure now is to add together all the percentage reactances and find their equivalent value. Reference should be made to Figure 15. It will be found that this value is approximately 214%.

In order to find the short-circuit voltamperes at the feeder, the following expression is used:

$$\text{Short-circuit MVA} = \frac{\text{Base MVA} \times 100}{\%X}$$

$$= \frac{150 \times 100}{214} = 70 \text{ MVA}$$

$$\text{Thus the short-circuit current } (I_f) = \frac{\text{Short-circuit MVA}}{\text{Line voltage}}$$

$$= \frac{70 \times 10^6}{\sqrt{3} \times 6.6 \times 10^3} = 6124 \text{ A}$$

Example 2

Figure 16 shows the connections at a factory substation. Each transformer has a rating of 500 kVA and a reactance of 8 per cent. The reactor installed to limit fault current has a rating of 300 kVA with a 10 per cent reactance.

(a)   Express the per cent reactance of the reactor to a basis of 500 kVA.

(b)   If the circuit breaker at B is open and the circuit breakers at points A and C are all closed, what is the prospective fault MVA occurring at the point F in the event of a fault?

(c)   If the circuit breaker at B is closed, but those at positions marked A are opened, what is the new prospective fault MVA at F?

CGLI 1979 (Modified)

(a)   Reactor $\%X = \dfrac{10 \times 500}{300} = 16.67\%$

(b)   With reference to Figure 17(a) it will be seen that the reactor is in series with one of the transformers and that the equivalent reactance is a parallel circuit. This is found to be:

$$\frac{8 \times 24.67}{32.67} = 6\%$$

Therefore prospective fault voltamperes

$$= \frac{0.5 \times 100}{6} = 8.33 \text{ MVA}$$

(c)   With the reactor out of circuit, the equivalent percentage reactance becomes 4%. Therefore prospective fault voltamperes increases to 50/4 = 12.5 MVA.

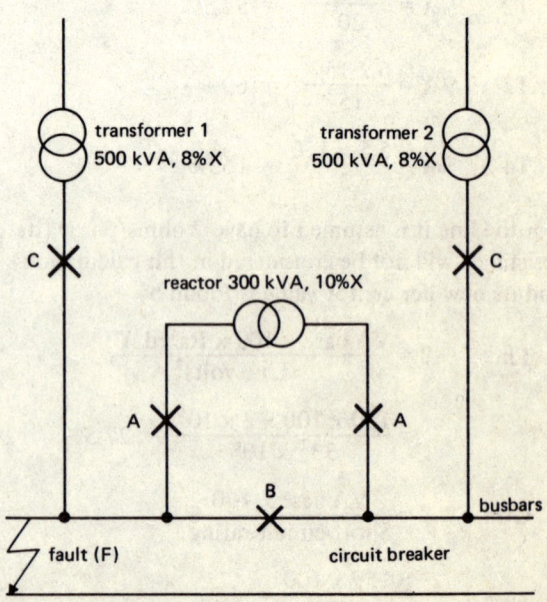

Figure 16

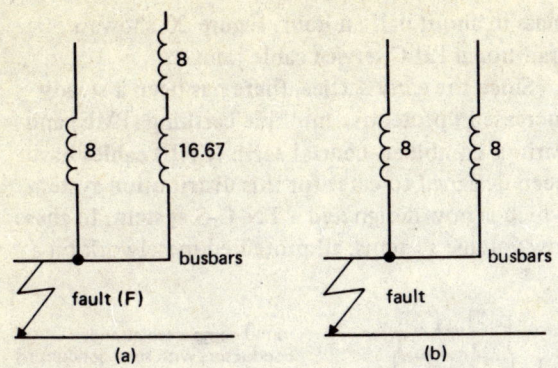

Figure 17

This example illustrates the importance of incorporating a reactor in the system. On three-phase systems at 415 V it is usual to ignore the high voltage system impedance and the prospective fault current is then the phase voltage divided by the sum of the phase impedances, that is

$$\text{Fault current } (I_f) = \frac{\text{phase voltage}}{\text{sum of phase impedances}}$$

$$= \frac{V_p}{Z_t + Z_m + Z_i}$$

where Z_t is the impedance of the transformer winding referred to the low voltage side

Z_m is the impedance per phase of the mains

Z_i is the impedance per phase of the installation

Example 3

It is required to find the prospective fault current from a delta–star 1000 kVA transformer and 50 m of 300 mm² aluminium cable. The following information is available:

	R	X
1000 kVA transformer	0.00219	0.00863
50 m of 300 mm² Al.	0.005	0.00123
	0.00719	0.00986

Phase impedance $(Z) = \sqrt{(R^2 + X^2)}$

$$= \sqrt{(0.00719^2 + 0.00986^2)}$$

$$= 0.0122 \ \Omega$$

Fault current $(I_f) = \dfrac{240}{0.0122} = 19.7 \text{ kA}$

On single-phase systems at 240 V it is often better to apply symmetrical component techniques. In practice, zero phase sequence values for cables are rarely available and the method often used is to sum the phase and neutral/protective conductor impedances.

Example 4

Find the phase-neutral fault current in the system described in Example 3 above having the following information:

	R	X
1000 kVA transformer	0.00219	0.00863
2 x 50 m, 300 mm² Al.	0.01	0.00246
	0.01219	0.01109

Phase neutral impedance $(Z) = \sqrt{(0.01219^2 + 0.01109^2)}$

$$= 0.01648 \ \Omega$$

Fault current $(I_f) = \dfrac{240}{0.01648} = 14.6 \text{ kA}$

Figure 18 provides more details of the supply network elements, starting from the generating source to a point on the low voltage distribution system feeding

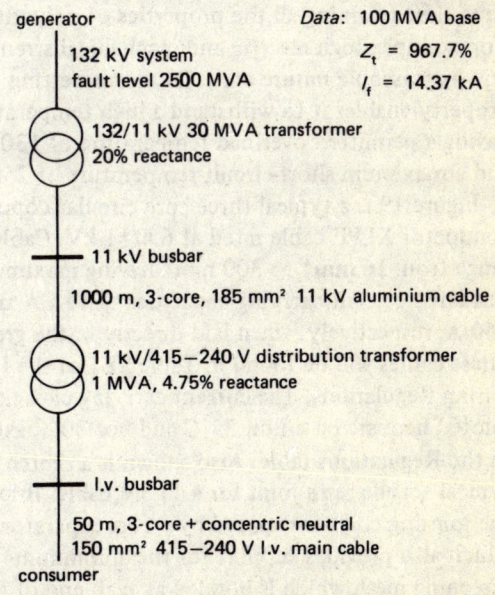

Figure 18 *Network elements determining short-circuit level when three-phase symmetrical fault occurs*

a consumer's premises. In this example a common base of 100 MVA is chosen and the short-circuit level found to be 14.37 kA. If the consumer was connected nearer the distribution transformer, then the level would rise, but if a smaller distribution transformer was employed, say 500 kVA, then the short-circuit level would be reduced. In practice, some supply authorities quote a value of prospective short-circuit current around 16 kA on low voltage mains. It is not possible for them to guarantee a level lower than this in view of supply network development and other factors.

Power cables

One of the most significant changes seen over the years in the cable industry has been in the use of cross-linked polyethylene (XLPE) as an insulant for cable conductors. While PVC cables have tended to supersede paper/lead cables in terms of handling and jointing in general, their continuous operating temperature is only 70°C compared with XLPE which is 90°C.

XLPE is formed by the vulcanization or cross linking of polyethylene, transforming it from a thermoplastic to a thermoset material while at the same time retaining all the properties of polyethylene (for example, high electric and mechanical strength, non-hygroscopic nature etc.). Its thermosetting property enables it to withstand a high temperature having a permitted overload temperature of 130°C and a maximum short-circuit temperature of 250°C.

Figure 19 is a typical three-core circular copper conductor XLPE cable rated at 6.4/11 kV. Cables range from 16 mm² to 300 mm² having maximum sustained current carrying capacities of 115 A and 540 A, respectively, when laid directly in the ground. These cables will be found in Table 9E1 of the IEE Wiring Regulations. The current carrying capacities quoted here are based on 25°C and not 30°C as given in the Regulations table. Also shown is a sketch of a typical acrylic resin joint for a 11 kV cable. Briefly, the jointing cores are spaced by a core separator which also provides support for the aluminium screening mesh which is bonded at each end of the joint to the earth metal components of the cable. The joint is enclosed in a moulded plastic box which is then filled with acrylic resin which cures into a solid

mass in about half an hour. Figure 20 shows a traditional PILC service cable joint.

Since the early sixties, there has been a steady increase in protective multiple earthing (PME) and various combined-neutral earth (CNE) cables have been designed to cater for this distribution system which is now designated a TN–C–S system. In these low voltage systems, all protected metalwork on a

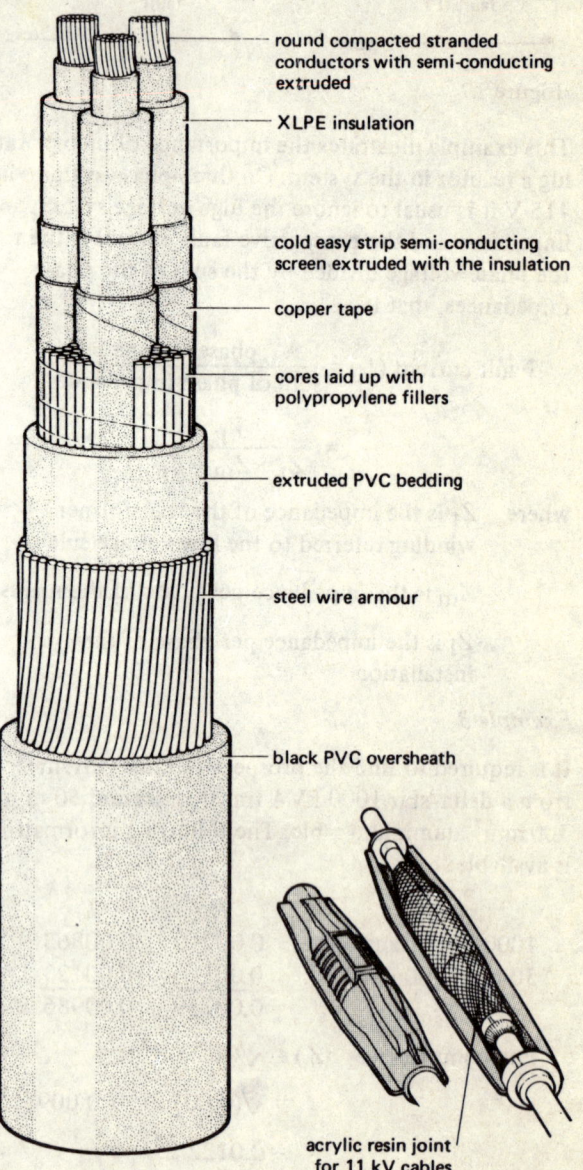

round compacted stranded conductors with semi-conducting extruded

XLPE insulation

cold easy strip semi-conducting screen extruded with the insulation

copper tape

cores laid up with polypropylene fillers

extruded PVC bedding

steel wire armour

black PVC oversheath

acrylic resin joint for 11 kV cables

Figure 19   *Three-core 6.4/11 kV XLPE power cable*

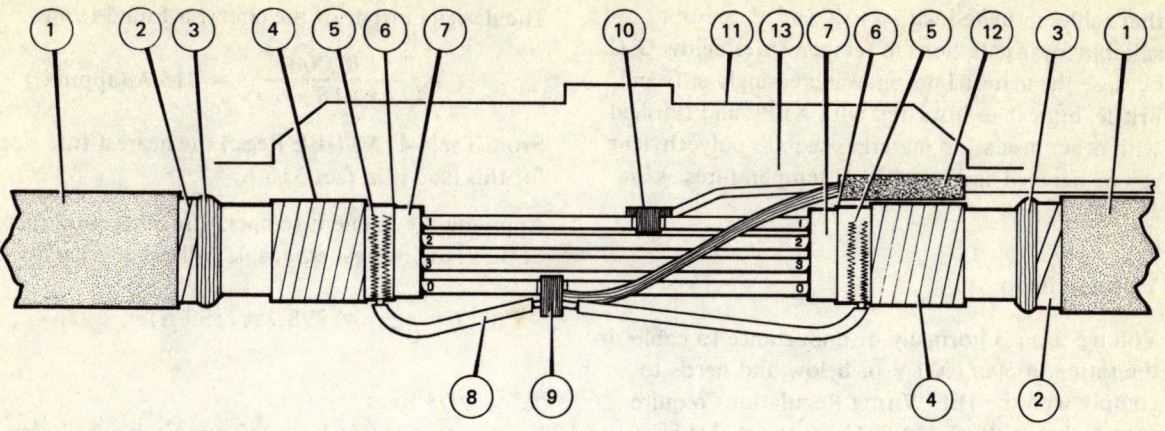

Key

1 Jute serving
2 Steel tape armour
3 Armour plumbed to lead sheath
4 Two layers of PVC tape bound over sheath
5 35 mm² copper continuity bond plumbed to lead sheath
6 Lead cut
7 Belt papers

8 35 mm² copper continuity bond
9 Neutral/earth soldered half ferrule connection (tape insulation not shown)
10 Phase connection bound and soldered (tape insulation not shown)
11 Joint mould top half only shown
12 Service cable sheath rasped for 80 mm
13 Polyurethane resin filling

Figure 20   *Sketch of service joint in four-core PILC cable*

consumer's premises is connected by protective
conductors to the neutral conductor of the supply
cable at the intake position. In order to avoid the
potential of the neutral rising to an unacceptable level
the system is earthed at the substation and earthed at
the end of each cable main as well as other points
along cable routes. This is to ensure that the maximum
resistance between neutral and earth does not exceed
20 Ω.

   Service cables for use on TN–C–S systems may be
single-phase or three-phase with both having a concentric neutral/earth (that is, a PEN conductor). In
practice, most single-phase service cables are PVC
insulated or XLPE insulated having phase circular
stranded copper conductors or circular solid
aluminium conductors with concentric copper
neutrals. Three-phase service cables will be XLPE
insulated with copper or aluminium conductors and
either copper concentric neutrals or aluminium waveform wire concentric neutrals. Figure 21 is a sketch
of a *waveformal* service cable. This cable and other
service cables used for TN–C–S systems are described
in Volume 2 of this series. It should be mentioned

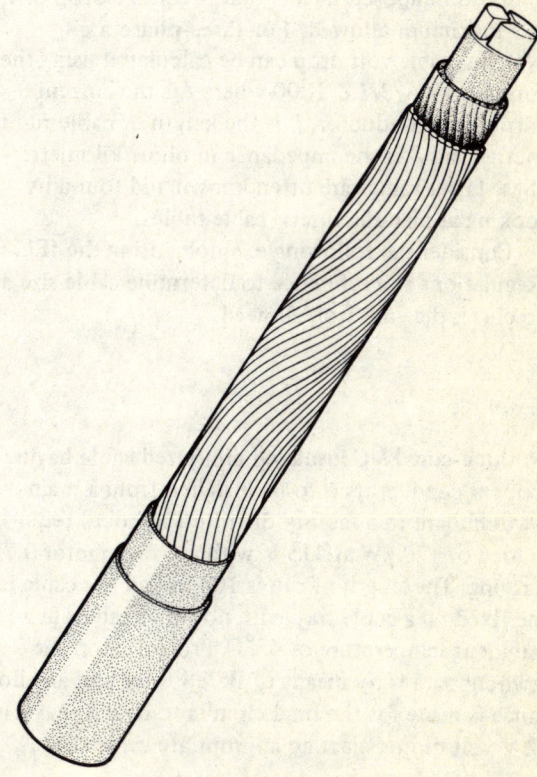

Figure 21   *Waveformal cable*

that cables insulated with a PVC sheath are not suitable for installations in temperatures below O°C because the material becomes increasingly stiff and brittle, but cables insulated with XLPE and finished with other insulation materials such as polyethylene can be handled and installed at temperatures as low as –40°C.

Voltage drop

Voltage drop is normally of importance to cables in the rating of 600/1000 V or below and needs to comply with the IEE Wiring Regulations requirements, that is, Reg. 522-8. Here, the designer of the installation has to consider the safe functioning of the electrical equipment. The regulation states that final circuits protected by overcurrent devices not exceeding 100 A will meet this requirement if the voltage drop from the origin of the circuit to any other point does not exceed 2.5% of the nominal voltage at the design current. For domestic installations, the 2.5% limitation imposed means that a voltage drop of 6 V is a maximum allowed. For three-phase a.c. systems, cable volt drop can be calculated using the formula $V = \sqrt{3}ILZ/1000$ where I is the current carried per conductor, L is the length of cable run in metres and Z is the impedance in ohms/kilometre – these latter values are often known and found by looking at manufacturers' cable tables.

Consider the following example, using the IEE Regulations as a reference to determine cable size and to check the volt drop allowed.

Example

A three-core PVC insulated armoured cable having copper conductors is to be installed from a main switchboard to a factory distribution board requiring a load of 170 kW at 415 V with a power factor 0.75 lagging. The length of run is 100 m and the cable is to be fixed on a cable tray with no other cables in an ambient temperature of 45°C. Protection in the switchboard is by means of BS 88 fuses and an allowance is made for the final circuits to have a maximum 2 V volt drop. Select an appropriate cable size.

The maximum permitted volt drop on the feeder cable is 10.375 – 2.0 = 8.375 V

The design current of the circuit is found from:

$$I = \frac{170{,}000}{\sqrt{3} \times 415 \times 0.75} = 315 \text{ A (approx.)}$$

From Table 41A2 (IEE Regs.) the nearest fuse size for this load is in fact 315 A.

Applying the ambient temperature correction factor of 0.79, the nearest size cable will need to carry:

$$I = \frac{315}{0.79} = 398.7 \text{ A (399 A)}$$

Table 9D3 Regs.
From columns 1 and 5 a 240 mm² cable is chosen having a volt drop/A/m of 0.2 mV.

Checking its volt drop to be less than the permitted:

$$V = \frac{100 \times 315 \times 0.2}{1000} = 6.3 \text{ V}$$

This cable is satisfactory.

It is interesting to point out here that if an XLPE cable was chosen (see Table 9E1 of the Regs.) instead of the PVC cable, a 150 mm² cable would satisfy current requirements, but it would not satisfy voltage drop requirements. Despite this, an XLPE 185 mm², which satisfies both conditions, could be used.

Earthing

Several regulations pertaining to the 1937 Electricity Supply Regulations require a supply undertaker, that is, an Electricity Board, not to commence or continue to give a supply of electricity to a consumer unless all metalwork, other than circuit conductors, whether enclosing, supporting or associated with the installation, is where necessary to prevent danger connected with earth.

An electricity board, under the above requirements is required, unless otherwise allowed, to earth the low voltage supply of the system at one point, usually the star point of the secondary winding of the supply distribution transformer at the substation. Where, however, a system of protective multiple earthing is used, the statutory requirement is modified to allow the system to be earthed at more than one point. By earthing the supply as suggested, the whole system is

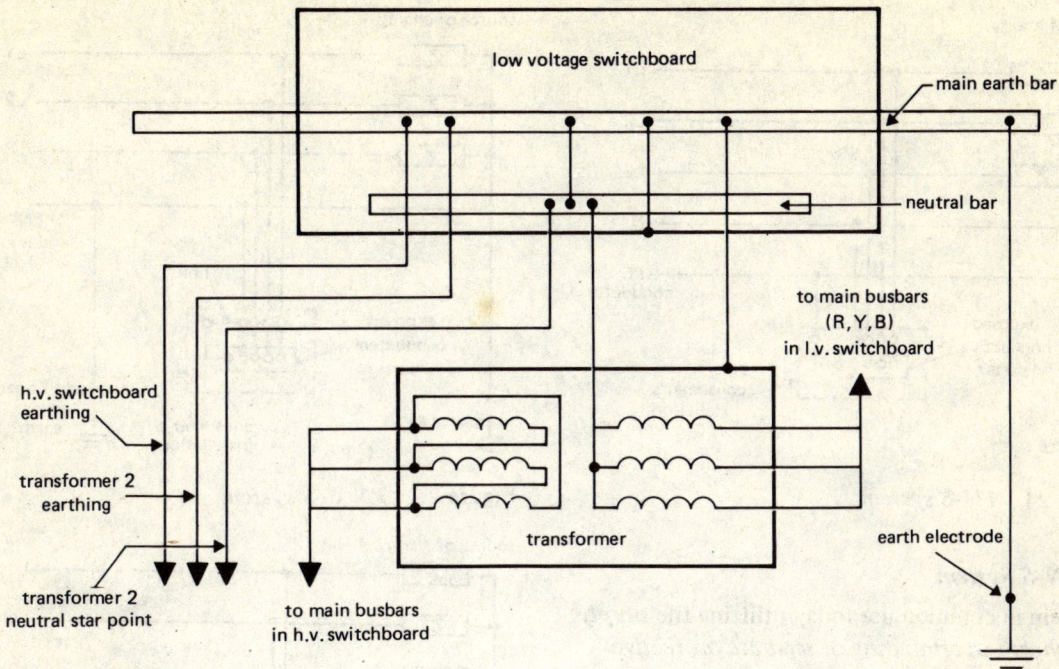

Figure 22   *Substation earthing*

tied to the potential of the general mass of earth. This not only ensures the potential on each conductor is restricted to values consistent with the insulation applied to them but it also prevents the passage of current through the earth under normal conditions and reduces problems associated with electrolysis and interference with communication circuits. Figure 22 is typical of substation earthing.

An efficient and effective earthing system is an essential ingredient of our a.c. supply in that it provides a low impedance and current escape path to enable protective devices in circuits to operate. Under requirement 413-3 of the IEE Wiring Regulations, the characteristics of the protective devices for automatic disconnection, the earthing arrangements and the relevant impedances of circuits need to be co-ordinated, so that during an earth fault the prospective shock voltage within the equipotential zone is limited in magnitude and duration time. The limiting values of impedances are given in Appendix 7 and 8 of the IEE Wiring Regulations.

Classification of present day earthing systems is given in Appendix 3 of the IEE Wiring Regulations. Briefly they are:

(a) *TN–C System*

A system where the neutral and protective conductor are combined to form a PEN conductor, such as *earthed concentric wiring*. Figure 23 shows this system which is restricted in its use to installations not connected directly to the public supply, for example, private generating plant.

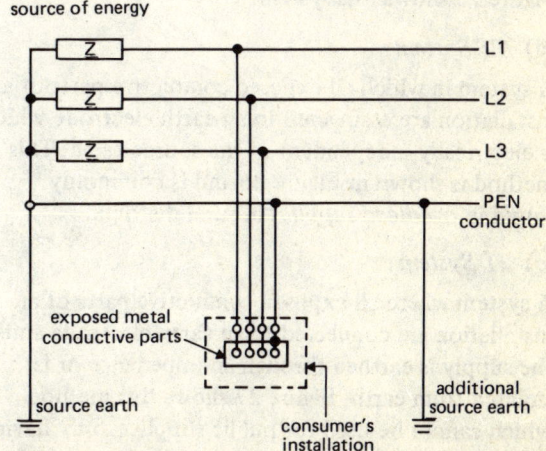

Figure 23   *TN–C system*

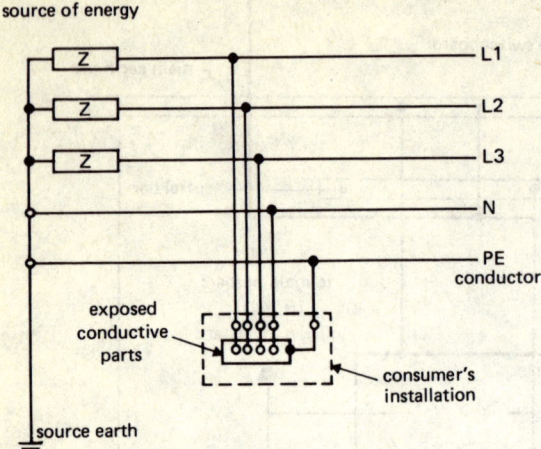

Figure 24   *TN–S system*

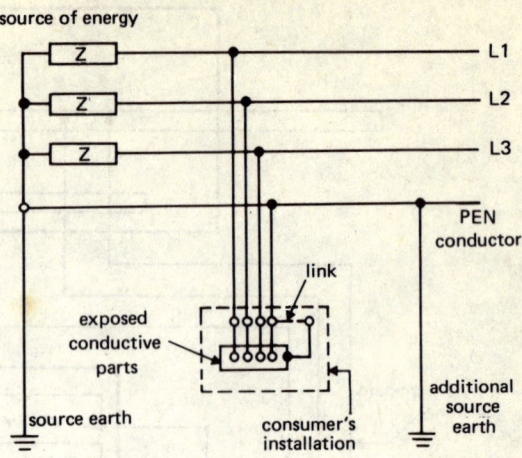

Figure 25   *TN–C–S system*

(b) *TN-S System*

A system in common use today utilizing the *supply cable sheath* or *armouring* or *separate protective conductor* (PE). In this method all exposed conductive parts of the installation are connected to the PE conductor via the main earthing terminal. Figure 24 shows this system.

(c) *TN-C-S System*

A system known as *protective multiple earthing* in which the neutral conductor and protective conductor are combined as a PEN conductor or combined neutral earth conductor (CNE conductor). A separate protective conductor is utilized within the installation. Figure 25 shows this system.

(d) *TT System*

A system in which all exposed conductive parts of an installation are connected to an earth electrode which is electrically independent of the source earth. This method is shown in Figure 26 and is commonly found in *overhead supplies* in rural districts.

(e) *IT System*

A system where all exposed conductive parts of an installation are connected to an earth electrode and the supply is earthed through an impedance or is isolated from earth. Figure 27 shows this method which cannot be used for public supplies, only having special use in *mines and quarries* and other *special applications*.

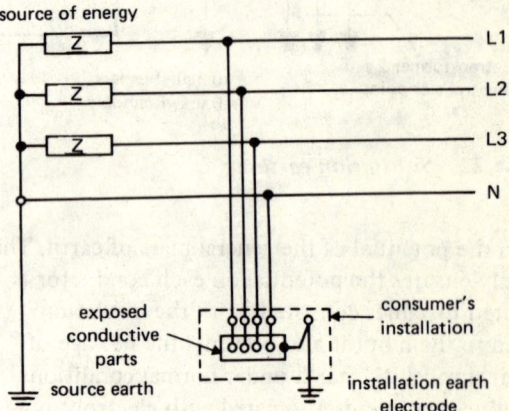

Figure 26   *TT system*

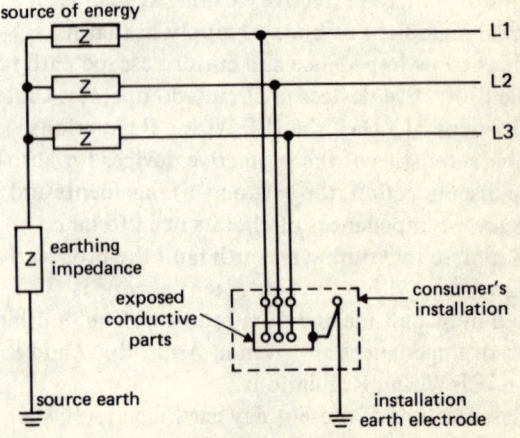

Figure 27   *IT system*

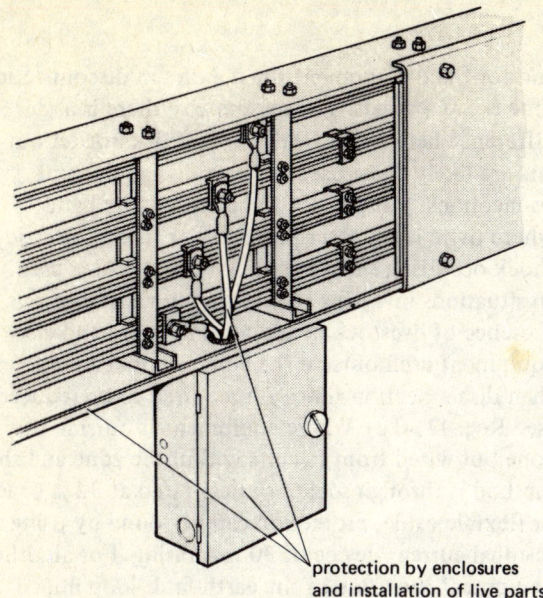

protection by enclosures
and installation of live parts

Figure 28   *Methods of protection against direct
contact*

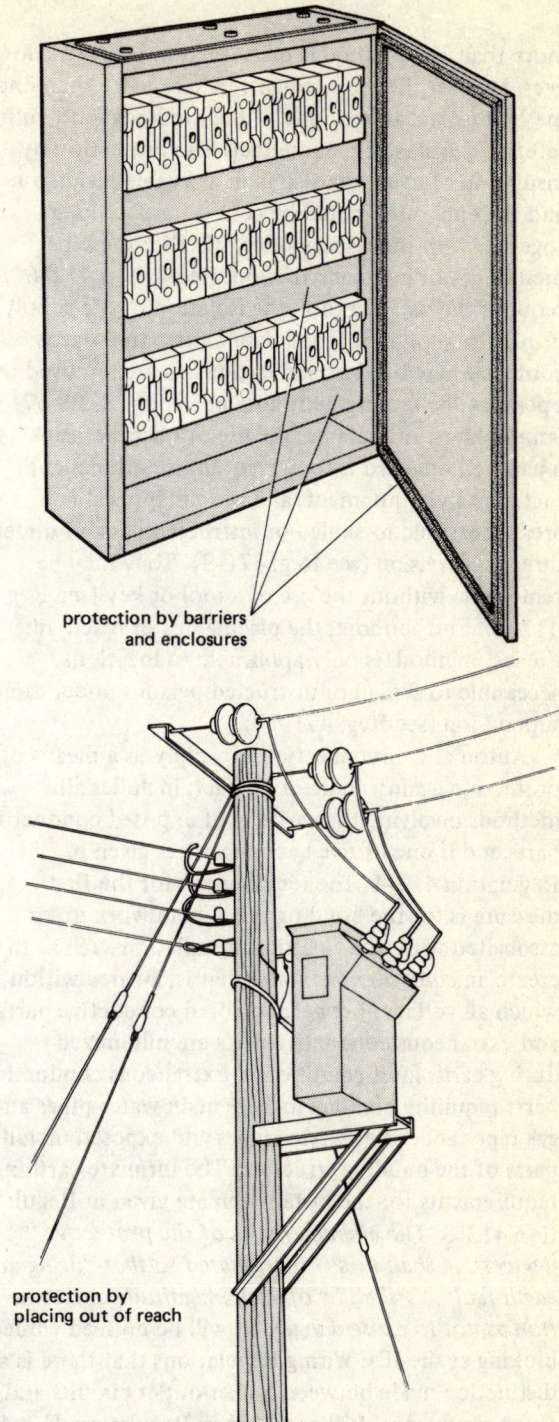

protection by barriers
and enclosures

protection by
placing out of reach

The two most common systems operating in this
country are TN–S and TN–C–S with the latter system
being used more and more on new installations. In
these systems the earth loop impedance will normally
be low enough to permit fuses and miniature circuit
breakers to be used for shock risk protection, but if
this is not so, a residual current device must be used.
These devices are now available in a wide range of
sensitivities and for a specific installation their
selection has to be based on the rated operating
current of the device multiplied by the earth loop
impedance and must not exceed 50. It should be
pointed out that a residual current device must not be
used in circuits incorporating a PEN conductor,
used throughout the system. In other earthing
systems where high earth fault loop impedances
exist, use is made of fault-voltage operated devices
but these are becoming obsolete.

The 15th Edition of the Wiring Regulations
recognizes two kinds of risk associated with electric
shock, namely, by *direct contact* with metalwork that
is live in normal service and by *indirect contact* as a
result of contact with metal work that could become
live under fault conditions. The former risk requires
protection by insulation of live parts, barriers or
enclosures, obstacles and placing out of reach. The
latter risk requires automatic disconnection of the
supply; the use of Class II equipment or equivalent

insulation; non-conducting locations; earth free local
equipotential bonding and electrical separation.
Figure 28 illustrates several methods of protection
against direct contact and one should appreciate that

more than one method is often used in practice. However, to meet the requirements, where more than one method exists, at least one of the methods must fully safeguard against direct contact. The protection by insulation of live parts also includes cable insulation and the enclosures could be conduit or trunking together with their accessories, whether these be metal or plastic in construction. Regulation 412–6 requires the use of a tool, the isolation of the supply or provision of an intermediate barrier to prevent contact where barriers and enclosures are removed or opened – the requirement does not apply to BS 67 lampholders. In practice, the use of obstacles such as a handrail or guard is to prevent unintentional contact near live equipment, and is only applicable in areas accessible to skilled or instructed persons under direct supervision (see Reg. 471–9). They may be removable without the use of a tool or key (see Reg. 412–7). Furthermore, the placing out of reach, like the last method, is only applicable to locations accessible to skilled or instructed persons under direct supervision (see Reg. 471–10).

Automatic disconnection of supply as a means of protection against indirect contact, includes all methods involving the earthing of exposed conductive parts and is one of five basic measures given in Regulation 413–1. The requirement for the first measure is for the bonding of all metalwork in, or associated with, the electrical installation such as to create an *equipotential zone* (that is, an area within which all voltages between exposed conductive parts and extraneous conductive parts are minimized during earth fault conditions). Extraneous conductive parts requiring bonding include main water pipes and gas pipes and other service pipes and exposed metallic parts of the building structure. The ultimate earthing requirements for the installation are given in Regulation 413–3 *'The characteristics of the protective devices . . . shall be so co-ordinated so that during an earth fault . . . shall be of such magnitude and duration as not to cause danger.'* It will be noticed while looking at the IEE Wiring Regulations that there is a distinction made between socket-outlet circuits and fixed equipment. It is specified in Regulation 413–4 that in order to satisfy the previous regulation, the earth fault loop impedance at every socket outlet must be sufficiently low enough to allow a disconnection of the supply to occur within 0.4 seconds

and for fixed equipment the maximum disconnection time is 5.0 seconds. The reason why there is a difference here stems from the fact that socket outlets are likely to be used for portable equipment connections, those likely to be gripped by hand where there is a greater posibility or risk of electric shock occurring since a trailing flex or lead is used. In situations involving reduced body resistance, presence of livestock in and around farms and also equipment used outside the equipotential zone, then disconnection times are required to be reduced (see Reg. 471–12). Where equipment is outside the zone but wired from circuits within the zone and the method is through socket outlets rated at 32 A or less or flexible cable, protection can be found by using a residual current device of 30 mA rating. For all other equipment the value of the earth fault loop impedance has to be such that disconnection of the supply occurs within 0.4 seconds. Where however protection is afforded by overcurrent protective devices, then the maximum earth fault loop impedance given in Tables 41A1 and 41A2 of the IEE Regulations should be adhered to, these being based on 240 V and negligible earth fault impedance. For other values of earth fault loop impedance refer to the note at the foot of the Tables.

The requirements of the IEE Wiring Regulations for circuit protective conductors with regard to thermal constraints are specified in Section 543 where Regulation 543–1 states two alternative methods of selecting the size of this conductor. One is the calculation of the c.s.a. using the adiabatic equation applicable for disconnecting times not exceeding 5 seconds; the other relates to the c.s.a. of the circuit protective conductor to the associated phase conductor as given in Table 54F of the Regulations.

It is a requirement of Regulation 313–1 that the earth fault loop impedance (Z_s) at the origin of the installation be ascertained and one of the involvements here is a knowledge of the external earth loop impedance (Z_e). Guidance of this value can be given by a supply authority for a proposed installation but can be measured if the supply is available. It has been mentioned elsewhere that for a TN–C–S system, Z_e is not likely to exceed 0.35 Ω – based on a 315 kVA distribution transformer and aluminium mains and service cable.

Consumers' distribution

For large consumers of electricity, the distribution system should be designed to provide as much flexibility as possible within the constraints allowed. There should be provision made for any known future extensions. Among other things there should be adequate spare capacity on main busbars and at sub-main distribution centres as well as on main and sub-main feeders. There should also be sufficient space allowed for additional switchgear and distribution equipment. The choice of supply voltage and type of distribution is often influenced by the building or buildings, their size, shape and situation, for example. In the design stages of the distribution system an assessment is required of the protective switchgear, fault levels, tripping characteristics and any emergency supply provision such as standby generating plant.

The distribution of electricity between the main intake and load centres may be either radial or ring systems. Where high voltage is employed for the main distribution system, the system should attempt to cover as much as is practicable with high voltage cables and keep the low voltage cables as short as possible; this is particularly important on premises with scattered buildings and multi-storey buildings with lateral mains on each floor. Where high voltage switchgear is provided by the supply authority, it may be unnecessary for the equipment to be duplicated since it is often possible to have an arrangement with the supply authority whereby the equipment can be shared. It is important for all electrical equipment to be capable of dealing safely with the maximum prospective fault energy to which it may be subjected, and, as pointed out earlier, this will depend on the arrangements of the supply distribution network.

Figure 29 is a typical arrangement of a high voltage *ring main* system having oil-immersed type switches and fused-switch protection for the transformers. A more selective system might employ directional protection using graded time-lag relays which bring about isolation to the faulty section of

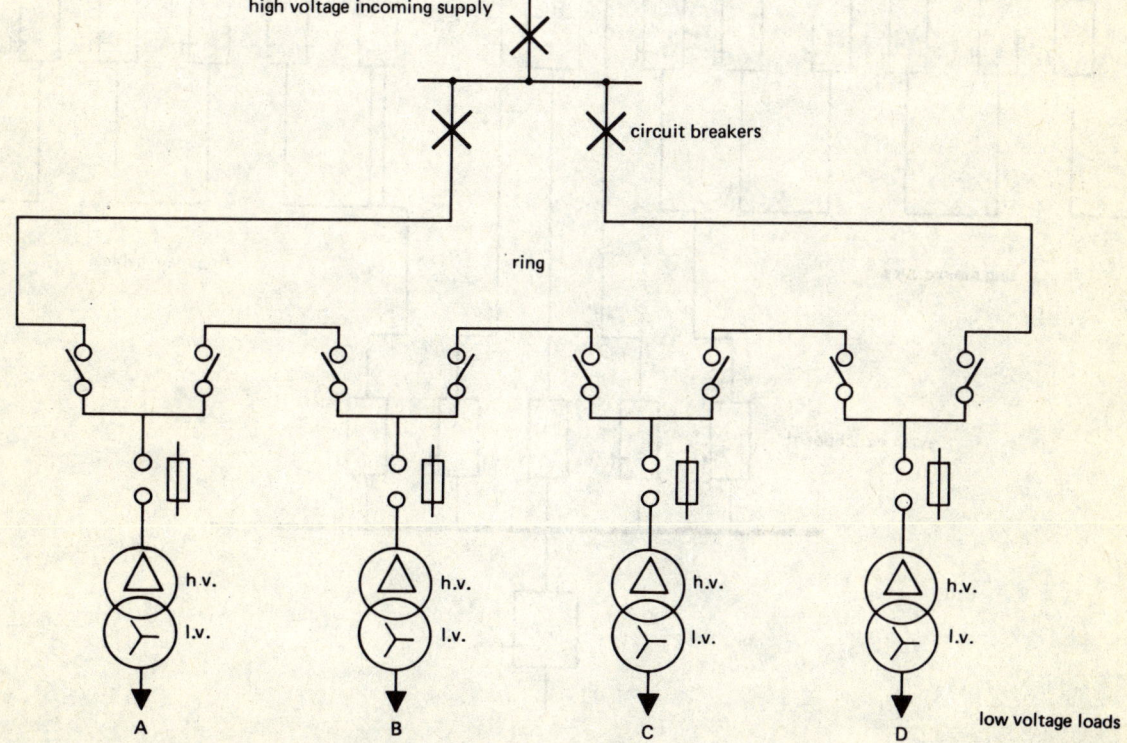

Figure 29   *High voltage ring main distribution system*

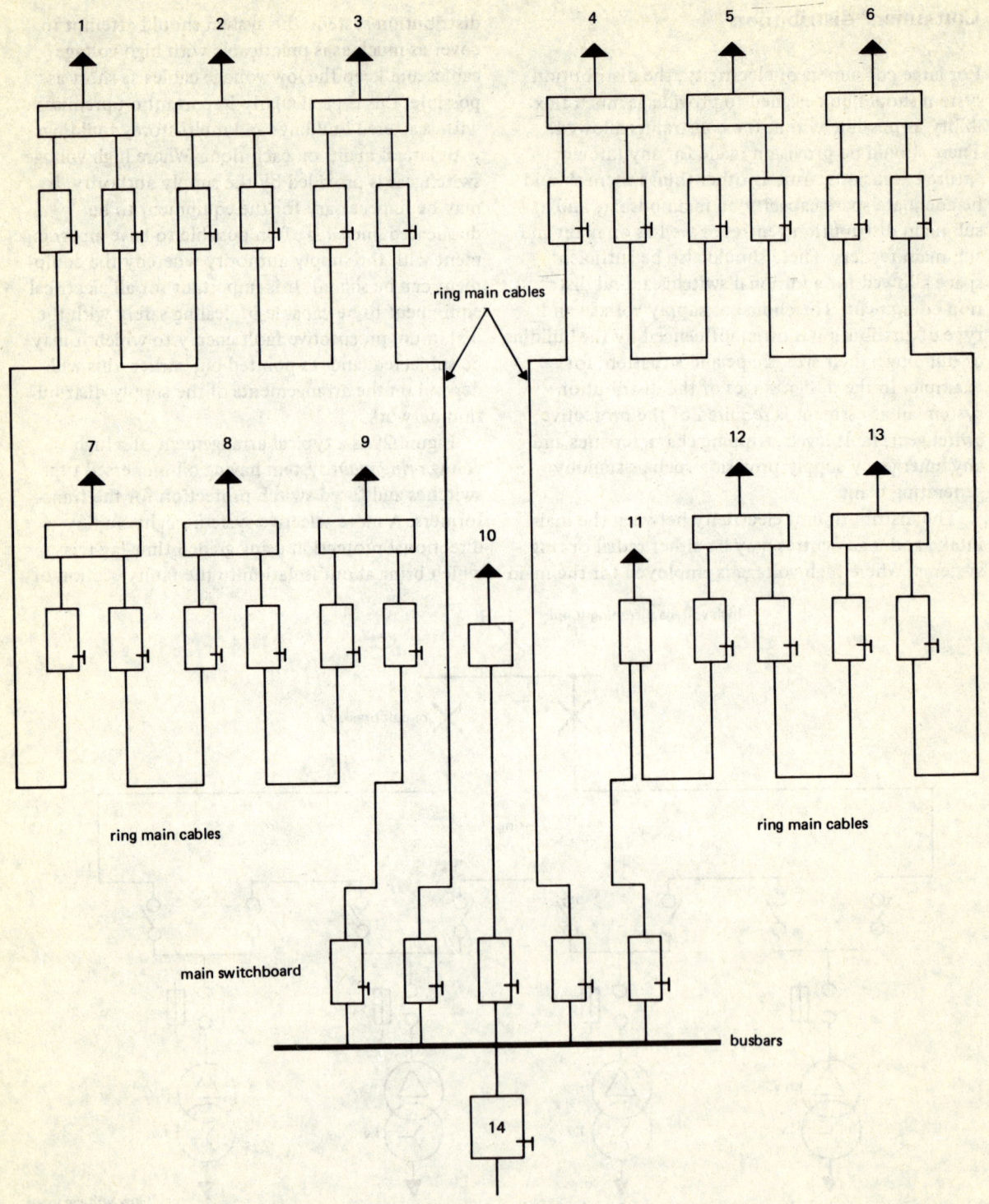

ring main cables

ring main cables

ring main cables

main switchboard

busbars

Figure 30   *Arrangement of switchgear for large hospital complex*

the ring. Figure 31 shows a high voltage *radial main* but with this distribution the isolation on fault is limited to the faulty section only. Where emergency standby power is required for essential services, this will be connected as a ring via the non-essential section of the switchboard when normal supply is off for maintenance or repairs. It is important to have key-interlock arrangements between the high voltage isolator and low voltage isolator to prevent paralleling of supplies.

In practice, it is quite likely to find high voltage switchgear being of the oil-break type with air-insulated busbar chambers being used. Oil switches are less costly than oil circuit breakers but are not suitable where time graded discrimination protection is preferred unless they are used in conjunction with other forms of excess current protection. With regard to high voltage transformers, oil-immersed types have a long history of reliability and should only be installed at ground level. Mineral oil in transformers could pose a possible fire risk and where there is a possibility of this occurring, askarel filled transformers are used. Askarel is a synthetic insulating liquid which is non-flammable and contains polychlorinated biphenyls (PCBs) which is most suitable for Class A insulation. While askarel transformers are delivered as sealed units, caution is required to prevent any spillage which could be harmful to the environment – any leaks should be reported to the manufacturer. Silicon liquid (*DC* 561) is now used.

In terms of low voltage distribution, switchgear should have adequate capacity for the estimated maximum loading as well as some spare capacity for future growth. Again, the type of distribution used will be influenced by the layout of the building and location of the loads. Cubicle type switchboards are not uncommon and are neater in appearance than unit type switchboards. Air-break switches complying with BS 861 and air-break circuit breakers complying with BS 4752 are typically employed. Circuit protec-

Key

1	Kitchen and recreation block supplies	— 2 x 500 A TPN fuse switches, fused at 450 A — 1 x 800 A, 4-bar busbar chamber
2	Male blocks (M1,M2,M3 and M4)	— 2 x 500 A TPN fuse switches, fused at 450 A — 1 x 800 A, 4-bar busbar chamber
3	Male occupation therapy and blocks M7 and M8	— 2 x 500 A TPN fuse switches, fused at 450 A — 1 x 800 A, 4-bar busbar chamber
4	Female occupation therapy and blocks F7 and F8	— 2 x 500 A TPN fuse switches, fused at 450 A — 1 x 800 A, 4-bar busbar chamber
5	Female blocks F9 to F17	— 2 x 500 A TPN fuse switches, fused at 450 A — 1 x 800 A, 4-bar busbar chamber
6	Sanitorium	— 2 x 500 A TPN fuse switches, fused at 450 A — 1 x 800 A, 4-bar busbar chamber
7	Female blocks F1 to F6 and nurses home	— 2 x 500 A TPN fuse switches, fused at 450 A — 1 x 800 A, 4-bar busbar chamber
8	Laundry block	— 2 x 500 A TPN fuse switches, fused at 450 A — 1 x 800 A, 4-bar busbar chamber
9	Administration block	— 2 x 500 A TPN fuse switches, fused at 450 A — 1 x 800 A, 4-bar busbar chamber
10	Workshops	— 1 x 300 A TPN fuse swich on main switchboard
11	Staff houses and doctors' quarters	— 6 way TPN feeder pillar
12	Farm buildings and gamma house	— as per blocks 1 to 9
13	Beta house	— as per blocks 1 to 9
14	Main switch position	— 1000 A TPN air circuit breaker

Notes: Main switchboard comprises four 500 A TPN fuse switches plus one 300 A TPN fuse switch.

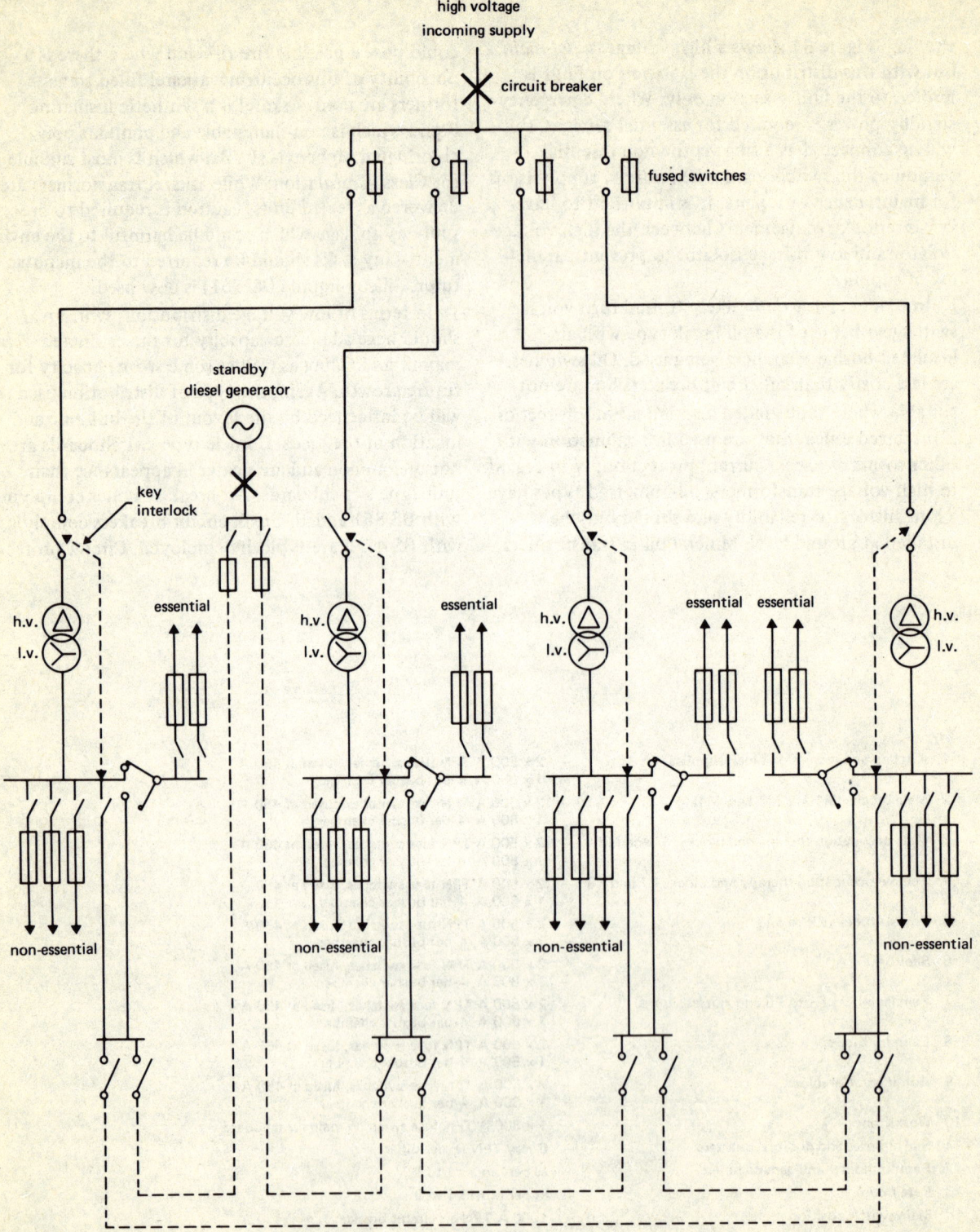

Figure 31   *High voltage radial distribution system with low voltage standby ring interconnector*

tion should be arranged to provide adequate discrimination and sub-distribution switchgear should be situated to minimize the length of final circuit wiring. It is important and also in line with the requirements of the IEE Regulations that switchgear and fuseboards should be clearly labelled to indicate the circuits they control. Figure 30 shows a low voltage ring main for a large hospital complex.

Revision exercise 2

1 (a) (i) What is meant by 'prospective short-circuit current'?
         (ii) How does it affect the required breaking capacity of switchgear?
    (b) In calculating short-circuit current, why is resistance generally ignored but reactance taken into account?
    (c) For a three-phase distribution system, write down expressions for calculating
         (i) short-circuit MVA
         (ii) short-circuit current
    (d) A 750 kVA, 11 kV/415 V three-phase transformer has a reactance of 5 per cent. If a short circuit of negligible impedance occurs on the secondary while rated voltage is applied to the primary, calculate the:
         (i) short-circuit MVA
         (ii) current

CGLI/C/81

2 (a) A transformer has a no-load and full-load secondary voltage of 500 V and 487 V respectively. Calculate the percentage regulation of the transformer.
    (b) Two three-phase generators and transformers are connected as shown in Figure 32. A short circuit across all phases occurs near the common busbars as shown. Calculate the short-circuit current.

3 (a) List *four* requirements regarding the electrical equipment found in a high voltage substation.
    (b) Make a line diagram of the high voltage and low voltage switchgear and transformer.
    (c) What method is used in a large capacity transformer to detect loss of oil and core winding fault?

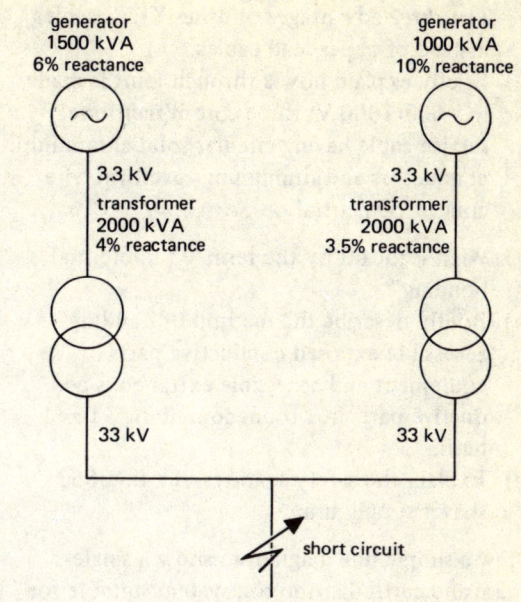

Figure 32

4 (a) Describe, using sketches, the construction of the following types of cable:
         (i) PVC insulated, split concentric for use on single phase distribution
         (ii) paper insulated, three-core cable with lead sheath and steel wire armour
    (b) Describe how to terminate *one* of the above cables into an isolator.
    (c) What precautions should be taken against corrosion when installing steel wire armoured cables underground?

CGLI/C/81

5 The cable feeding the distribution board serving an installation of fluorescent luminaires is to be a paper-insulated lead sheathed armoured type with copper conductors and is to be clipped to a wall in an area of ambient temperature of 40°C. The installation consists of 720, 65 W luminaires which operate with 10 per cent choke losses and a power factor of 0.82 lagging. If the length of run from the 415 V three-phase and neutral supply is 80 m calculate the smallest size of cable which may be used. The voltage drop on the final circuit wiring is a maximum of 2 V.

CGLI/C/82

6 (a) List *three* advantages of using XLPE cables instead of paper/lead cables.
   (b) Briefly explain how a through joint is made in a 600/1000 V, three-core *Waveformal* service cable having circular solid aluminium conductors and aluminium waveform wire concentric neutral.

7 (a) What is meant by the term 'equipotential bonding'?
   (b) Briefly describe the method of bonding accessible exposed conductive parts of equipment and accessible extraneous conductive parts in a room containing a fixed bath.
   (c) Explain the safety features of a BS 3052 shaver supply unit.

8 Draw a simple line diagram to show a single-phase and earth distribution system suitable for site lighting and for the use of portable and transportable tools on a building or construction site. On the diagram, show details of the incoming supply unit, main distribution unit, transformer unit and outlet units, and indicate any special features that may be required with regard to earthing, voltage, capacity of outlets and use of circuit breakers. (Assume that an incoming electrical supply of single-phase, 240 V, 50 Hz is available.)

CGLI/C/77

9 (a) State *three* advantages of a ring-main distribution over a radial distribution in an industrial installation.
   (b) Explain why miniature circuit breakers may not offer satisfactory discrimination under short-circuit conditions.
   (c) A factory substation is to be equipped with oil circuit breakers and it is decided to allow provision for future expansion of the factory premises.

Explain:
(i)   why all the oil circuit breakers required eventually might not be provided initially
(ii) what provision is normally made for future extension of the switchgear
(iii) what provision is normally made for additional distribution cables within the substation and across factory main access roads

CGLI/C/84

10 A 415/240 V TP & N lighting distribution board in an office block is fed by four 10 mm^2 single core PVC insulated cables (copper conductors) installed in PVC conduit. The length of run from the main switchboard is 85 m and the ambient temperature is 30°C. Assume that:

A 2.5 mm^2 protective conductor is employed. Protection at the main switchboard is afforded by 32 A fuses to BS 88 Part 2. The tested value of Z_e at the main switchboard is 0.4 Ω.

The impedance of final circuit phase and protective conductors does not exceed 0.5 Ω.

Establish the following:
(a) the prospective fault current due to an earth fault on the cables at the distribution board
(b) the disconnection time of the fault current in (a)
(c) by the use of Regulation 543-2, that the proposed protective conductor to the distribution board is satisfactory
(d) the maximum value of Z_S of the final circuits
(e) that the value of Z_S satisfies Regulation 413-4.

CGLI/C/84

chapter three
Special installations

After reading this chapter you will be able to:

1 State a number of requirements for electrical installations associated with:
   (a) hazardous areas
   (b) petrol filling stations and garages
   (c) agricultural and horiticultural premises
   (d) fire alarm systems
   (e) standby supplies
   (f) lightning protective systems
   (g) corrosive environments

2 Identify hazardous areas as zones and recognize methods of protection commonly associated with such zones.

3 State a number of precautions against explosions and burns occurring in hospital operating theatres.

4 Outline the risks associated with the wiring of petrol filling stations and garages.

5 Outline the risks associated with the wiring of agricultural and horticultural premises.

6 Recognize a number of alarm and triggering devices used in the wiring of fire alarm systems.

7 Know the methods used for providing emergency standby power to premises.

8 Describe the following terms used in association with a lightning protective system:
   (a) air termination network
   (b) earth termination network
   (c) down conductor
   (d) zone of protection

9 Distinguish between (a) a sacrificial anode system, and (b) an impressed current system with regard to cathodic protection.

10 Draw circuit diagrams of the following:
   (a) zener diode safety barrier
   (b) pipe-ventilated motor control
   (c) control and distribution of forecourt pump motors:
   (d) final circuit supplies to 110 V/16 A socket outlets;
   (e) mains-operated fence controller
   (f) simple fire alarm system
   (g) simple mains failure system
   (h) lightning protective system
   (i) cathodic protection schemes

Electrical apparatus in hazardous areas

There are probably hundreds of different types of flammable liquids, gases and vapours found in industrial processes today, particularly in the chemical and petrochemical industries associated with gas production and petroleum storage and distribution.

A hazardous area is defined in BS 5345, Part 1: 1976, as *an area in which explosive gas-air mixtures are, or may be expected to be, present in quantities such as to require special precautions for the construction and use of electrical apparatus.*

Most flammable substances form explosive mixtures with air only between certain concentration limits. A mixture below the lower explosion limit will be too lean and one above the upper explosion limit will be too rich. In both conditions an explosion will not result. The lowest temperature at which sufficient vapour is given off from a flammable substance to form an explosive gas-air mixture is called *flashpoint* and the term used to describe the lowest temperature whereby an explosive mixture can be ignited is called *ignition temperature.* Explosive mixtures differ considerably, for example, a mixture of air and town gas will ignite at 560°C whereas a mixture of air and petrol will ignite at 250°C.

Gases and vapours fall into well defined groups according to their ignition temperatures. Gases being distinguished between the mining industry (Group I) and surface industry (Group II), the latter being subdivided, see Table 1. Furthermore, a system has been created for classifying electrical apparatus according to its *maximum surface temperature,* that is, the highest temperature the apparatus can attain under practical conditions, including overload and fault conditions affecting the working conditions of the apparatus. The system of classification is called

Table 1

Gas	Apparatus group	Industry
Methane	Group I	Mining
	Group II	Surface
Propane	IIA	
Ethylene	IIB	
Hydrogen	IIC	

temperature class (T class). This and the gas grouping are marked on all electrical apparatus used in potentially explosive atmospheres. Table 2 illustrates the temperature class for the standard ranges of maximum surface temperature. Where the surface temperature is greater than that given in the table the actual value has to be shown. It should be noted that the reference ambient temperature is taken as 40°C.

Table 2

Class	Maximum surface temperature (°C)
T1	450
T2	300
T3	200
T4	135
T5	100
T6	85

Along with the gas grouping and temperature class are two other factors which are taken into consideration when selecting apparatus for use in hazardous areas; they are *type of protection* and *environmental conditions.* The first takes into account the *zone* risk or degree of probability that an explosive atmosphere will be present, whereas the second factor considers the construction of the electrical apparatus for its exposure to the weather, corrosion, ingress of liquids and solid foreign particles or even the effects of solvents and heat from other apparatus.

With regard to the zone risk, BS 5345: Part 1: 1976, defines three zones in order of severity:

Zone 0   In which an explosive gas-air mixture is continuously present, or present for long periods.

Zone 1   In which an explosive gas-air mixture is likely to occur in normal operation.

Zone 2   In which an explosive gas-air mixture is not likely to occur in normal operation, and if it occurs it will exist only for a short time.

It should be noted that BS 5345 does not deal with mining application or explosive processing manufacture and Part 10 of the code concerns combustible dusts.

In practice, however, defining zone boundaries calls for a committee to be set up of people who have expert knowledge of the potentially explosive atmospheres present. Zones can then be allocated, marked on a site plan of the project and a copy sent to the enforcement authority for approval and comment.

It is generally recognized that there should be no electrical apparatus installed in a Zone 0 area, but where this is completely unavoidable, only intrinsically safe ('ia') and special category ('s') equipment is allowed. These and other types of protection for use in Zone 1 and Zone 2 areas are briefly outlined below, see Table 3 and Figure 34.

Table 3

Zones	Type of Protection	
	UK	CENELEC
Zone 0	Ex ia	EEx ia
	Ex s	EEx s
	All above methods plus:	
	Ex ib	EEx ib
Zone 1	Ex s	EEx s
	Ex d	EEx d
	Ex p	EEx p
	Ex e	EEx e
	All above methods plus:	
Zone 2	Ex o	EEx o
	Ex q	EEx q
	Ex N	EEx n

1 Intrinsic safety (i)

'A protection technique based upon the restriction of electrical energy within apparatus and of interconnecting wiring, exposed to a potentially explosive atmosphere, to a level below that which can cause ignition by either sparking or heating effects. Because of the method by which intrinsic safety is achieved, it is necessary to ensure that not only the electrical apparatus exposed to the potentially explosive atmosphere, but also other electrical apparatus with which it is interconnected is suitably constructed.'

There are two classes of intrinsic safety, namely, 'ia' and 'ib'. The former is tested to a higher standard where safety is maintained with up to two faults, whereas the latter is tested with one fault applied. Both types are used in low energy circuits

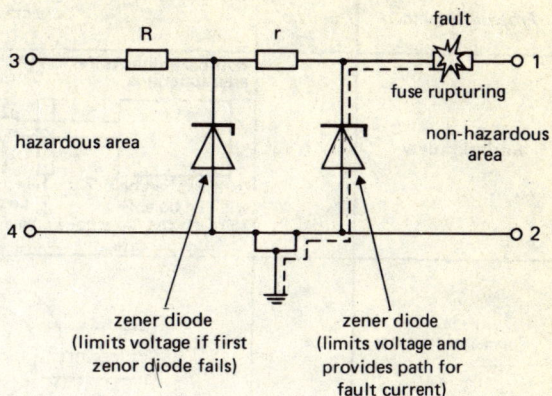

Figure 33 *Zener diode safety barrier circuit*

such as instrumentation and control circuits.

The method of intrinsic safety calls for the use of a *safety barrier* located in the non-hazardous area to protect the integrity of the circuit against a fault. The most common type used at the present time is called a *zener safety barrier* and is shown in Figure 33. It consists of a fuse to limit the maximum current that can flow through the zener diodes; two resistors, one to limit the maximum current entering the hazardous area (R), the other used to allow the diodes to be tested separately (r); also two forward-connected zener diodes which act to limit the transmitted voltage.

In normal operation, the voltage applied between 1 and 2 should be below the barrier working voltage since one does not want the zener diodes to conduct. The fuse will rupture when the voltage is too high as a result of a fault.

As a general guide, intrinsic safety circuits are ideal for low wattage electrical equipment in the order of a few watts. Figure 35 is a typical assembly of safety barriers in a standard weatherproof enclosure, Type N, approved for installations in low risk hazardous areas (Zone 2 areas).

2 Special protection(s)

'A concept that has been adopted to permit the certification of those types of electrical apparatus that, by their nature, do not comply with the constructional requirements specified for apparatus

Protection method	Symbol	Scheme	Applications
Intrinsic safety	i	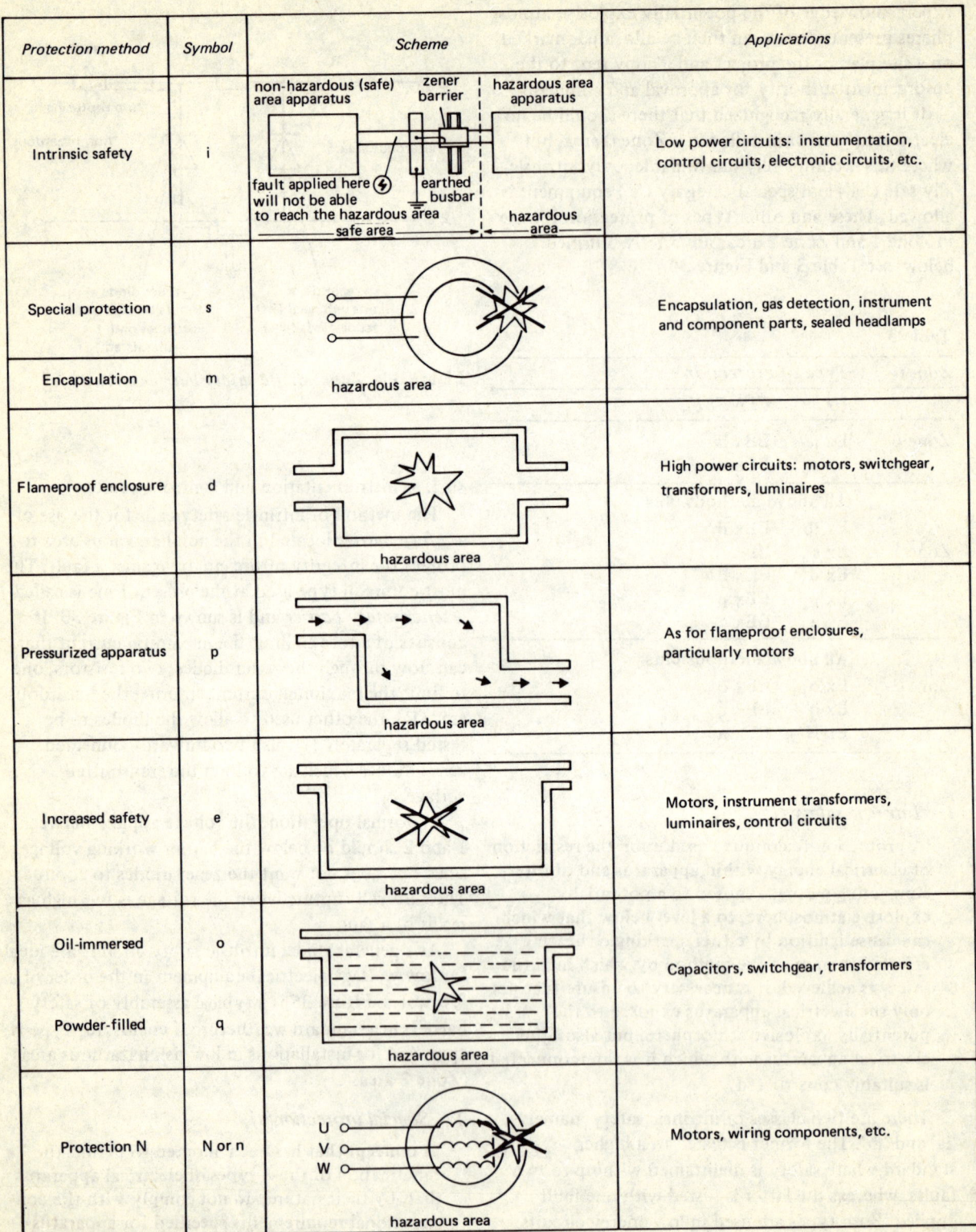	Low power circuits: instrumentation, control circuits, electronic circuits, etc.
Special protection	s		Encapsulation, gas detection, instrument and component parts, sealed headlamps
Encapsulation	m		
Flameproof enclosure	d		High power circuits: motors, switchgear, transformers, luminaires
Pressurized apparatus	p		As for flameproof enclosures, particularly motors
Increased safety	e		Motors, instrument transformers, luminaires, control circuits
Oil-immersed	o		Capacitors, switchgear, transformers
Powder-filled	q		
Protection N	N or n		Motors, wiring components, etc.

Figure 34

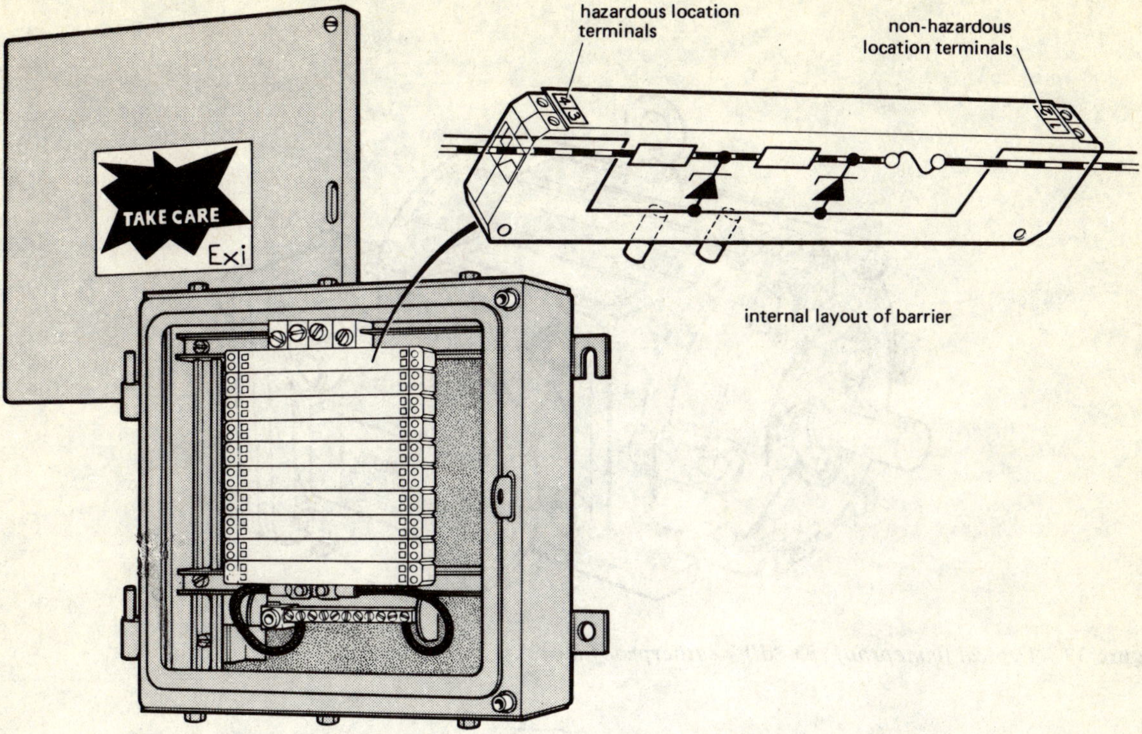

hazardous location terminals

non-hazardous location terminals

internal layout of barrier

TAKE CARE

Exi

Figure 35 *Safety barrier in weatherproof enclosure (by courtesy of Measurement Technology)*

with established types of protection, but which, nevertheless, can be shown, where necessary by tests, to be suitable for use in hazardous areas in prescribed Zones.'

An example of this method is shown in Figure 36. It will eventually be superseded by type of protection 'm' known as 'encapsulation'.

3   Flameproof enclosure (d)

'A method of protection where enclosures for electrical apparatus will withstand an internal explosion of flammable gas or vapour which may enter it, without suffering damage and without communicating the internal flammation to the external flammable gas or vapour for which it is designed, through any joints or structural openings in the enclosure.'

FIRE

BREAK GLASS PRESS HERE

KAC   KAC Alarm Co. Ltd
Alcester, England

Model K6
5Amp 240V AC
5Amp 30V DC Resistive
3Amp 30V DC Inductive

Figure 36 *Fire alarm manual call point having type 's' protection (by courtesy of KAC Alarm Co. Ltd)*

HISEAL
Exd

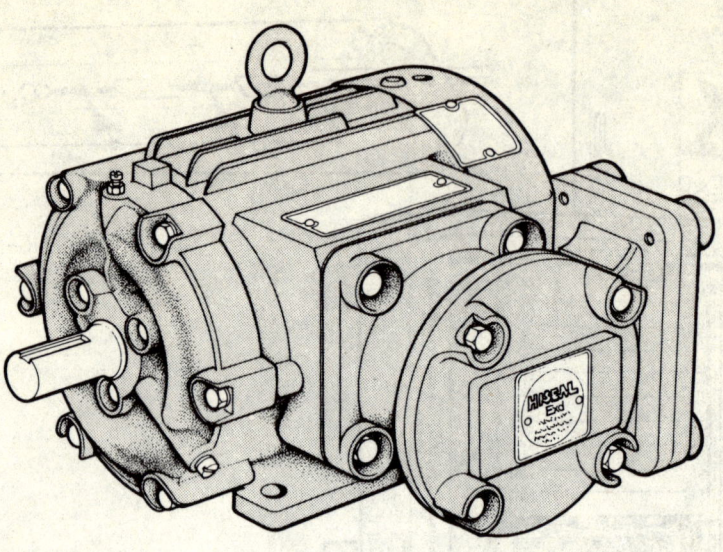

Figure 37 *Typical flameproof (Ex 'd') weatherproof motor*

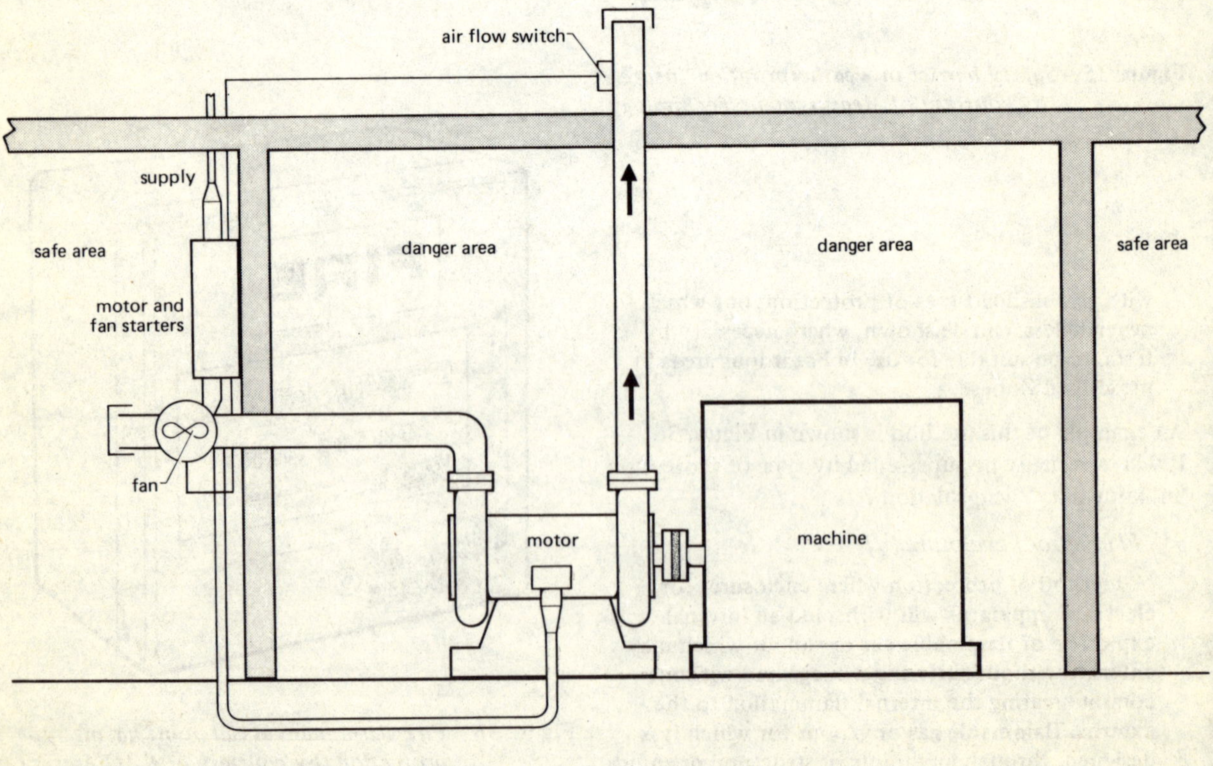

air flow switch
supply
safe area
danger area
danger area
safe area
motor and
fan starters
fan
motor
machine

Figure 38 *Pipe-ventilated motor*

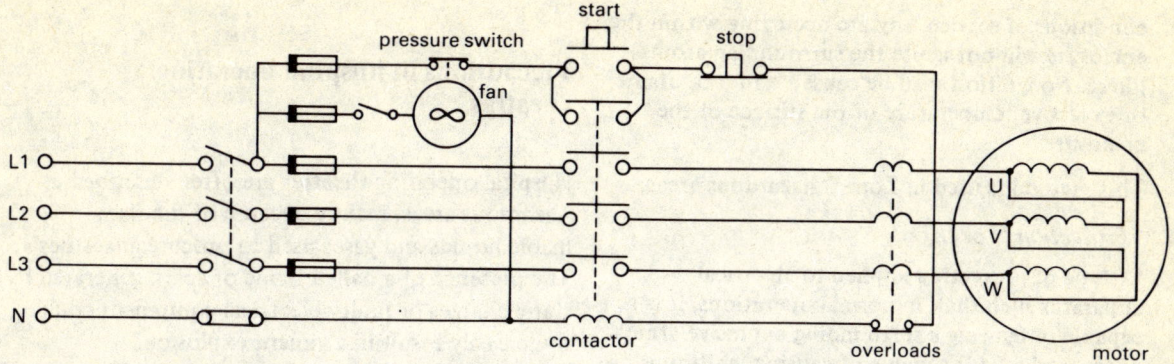

Figure 39 *Control circuit of pipe-ventilated motor*

Use of this method is often in high power circuits, such as motors, transformers, switchgear and electrical equipment. Figure 37 is an example of this method.

4 *Pressurized apparatus (p)*

'A type of protection by which the entry of a surrounding atmosphere into the enclosure of the electrical apparatus is prevented by maintaining, inside the said enclosure, a protective gas at a higher pressure than that of the surrounding atmosphere. The overpressure is maintained either with or without a continuous flow of protective gas.'

The application of this method is similar to that of (3) above. Additional precautions are needed in case of pressurization failure. Figure 38 illustrates this method using a pipe-ventilated motor. It will be seen that the motor circuit is provided with a pressure switch to monitor the flow of cooling air which is sited outside the flammable zone. In the electrical circuit diagram of Figure 39, it will be seen that a time delay has been incorporated to allow the motor to be purged with clean air before it is started.

5 *Increased safety (e)*

'A method of protection by which additional measures are applied to electrical apparatus to give increased security against the possibility of excessive temperatures and of the occurrence of arcs and sparks during the service life of the apparatus.'

The additional measures often use insulation materials with a high degree of integrity or enhanced

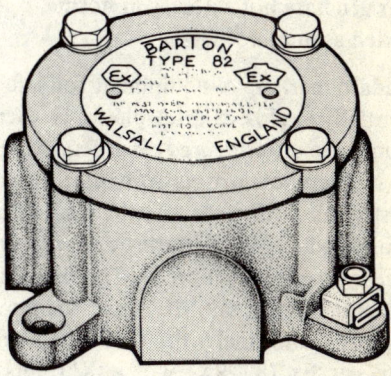

Figure 40 *'Barton' weatherproof junction box providing increased safety (Ex 'e') protection*

creepage and clearance distances. An example of this method is shown in Figure 40. The method is often used for induction motors and lighting fittings in Zone 1 hazardous areas.

6 *Oil-immersed apparatus (o)*

'A type of protection in which electrical apparatus or parts of electrical apparatus are immersed in oil in such a way that an explosive atmosphere which may be above the oil outside the enclosure cannot be ignited.'

7 *Powder-filled apparatus (q)*

'A type of protection in which the enclosure of the electrical apparatus is filled with a material in a finely granulated state so that, in the intended conditions of service, any arc occurring within the enclosure will not ignite the surrounding atomsphere. No ignition shall be caused either by flame or excessive temperature of the surface of the enclosure.'

This method is used in Zone 2 hazardous areas.

8   *Protection N or (n)*

'A type of protection applied to electrical apparatus such that, in normal operations, it is not capable of igniting a surrounding explosive atmosphere, and a fault capable of causing ignition is not likely to occur.'

Electrical apparatus designated this form of protection is suitable for use in Zone 2 areas only. The method is applied to apparatus which does not arc or spark or generate hot surfaces in normal operation. It is regarded as non-incendive.

Standard marking of equipment for safe selection in potentially explosive atmospheres is, therefore, based on three criteria: *area classification; gas classification, and temperature classification.* It will consist of a first symbol 'E' to denote the CENELEC standard, followed by 'Ex' denoting explosion protection, then the type of protection, say 'ia' then the apparatus gas group, say 'IIC', then the temperature class, say 'T4' and finally the BASEEFA certificate number, say 'Ex 78229X', where 'X' refers to special conditions. Electrical components are not marked with an 'X' but instead marked with the letter 'U'. Figure 41 is a flameproof adaptor identified as 'Exd IIC BAS 1111U'.

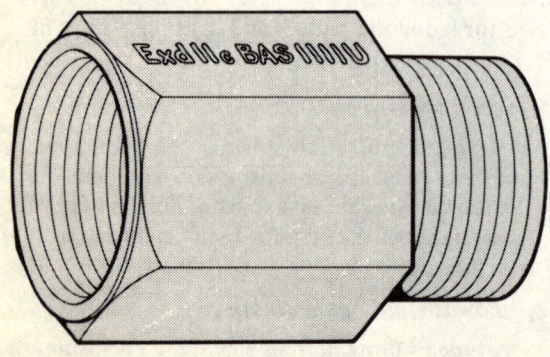

Figure 41   *Flameproof adaptor*

Precautions in hospital operating theatres

Hospital operating theatres are often described as hazardous areas, mainly because of the flammable liquids and gases used to procure anaesthesia. The presence of a naked flame or spark generated by static charges or faulty electrical equipment could quite easily result in a violent explosion.

Methods to overcome sparks generated by static charges have either been to fix chains on all metal apparatus so as to discharge any 'build-up' to earth or to fit antistatic rubber wheels to apparatus. Additional measures might include the removal of blankets before commencing anaesthesia, the non-use of ordinary rubber or plastic sheeting (unless the former is covered with a cotton material), the avoidance of wearing nylon clothing in the operating theatre, keeping the humidity of the atmosphere above 50 per cent and also the availability of antistatic footwear.

There is always a possibility of a spark being generated by one of the following:

(a) diathermy or cautery
(b) electric motor (for example, suction motor)
(c) light inside a bronchoscope or endoscope
(d) oxygen cylinders

In (a) above, a surgical diathermy is a means of producing heat by a high frequency current often used for sealing off bleeding points. By using a high frequency, electric shock, as commonly understood, is avoided. In practice, the main current is passed into a diathermy machine first before the high frequency is generated. A current is then concentrated through a small point, notably the active electrode of the diathermy, and body cells in contact with this electrode become destroyed. Current returns to the diathermy machine through a passive electrode which consists of a large lead plate covered with lint which has been moistened in hypertonic saline. The electrode has a large surface area so that the activity of the current is fanned out and not concentrated – this avoids the danger of burns occurring. It is important to see that the electrode is not cracked or that the electric flex does not become detached. The active electrode is controlled by a footswitch and must be conveniently placed for the surgeon.

With regard to (b) and (c), various methods are commonly used today, particularly the protective methods mentioned before, for example, 'Ex' equipment. In (d), however, it should be pointed out that the friction of gases escaping from oxygen cylinders can lead to heat being generated and the presence of oil or dust in the oxygen-rich atmosphere may be seen as a likely cause of explosion.

Petrol filling stations and garages

Under the provisions of the Petroleum (Regulations) Acts, 1928 and 1936, licensing authorities such as County Councils or District Councils, are required to make conditions for the licensing of premises used for keeping petroleum spirit. The electrical requirements and regulations concerning filling stations vary to some extent from area to area and licensing authorities publish their own requirements which are based on the Model Code of Principles of Construction and Licensing Conditions, Part I and II, published by the Home Office. Some local authority documents make reference to BASEEFA Certification Standard SFA 3002 but because of the considerable developments over the last ten years the Health and Safety Executive may soon publish a Code of Practice covering the construction principles and zoning of the forecourts/filling stations and also the construction, installation and maintenance of petrol pumps/dispensers will be covered by a British Standard.

The above mentioned Code and the Institute of Petroleum Electrical Safety Code give recommendations for the installation of electrical equipment in areas in which an inflammable atmosphere which could be ignited by an electrical source, may be present. These dangerous areas are again classified as zones and similar to the ones already described, they are:

Zone 1   An area within which any flammable substance whether gas, vapour or volatile liquid is processed, handled or stored, and where during normal operations an explosive or ignitable concentration is likely to occur in sufficient quantity to create a hazard.

Zone 2   An area within which any flammable or explosive substance, whether gas, vapour or volatile liquid although processed or stored, is so under conditions of control that the production (or release) of an explosive or ignitable concentration in sufficient quantities to constitute a hazard is only likely under abnormal conditions.

Safe areas are those which do not fall within the above but where a location apparently falls within the two zones, that location is to be regarded as being in the more hazardous zone.

The hazardous area of a forecourt is classified as Zone 2 and extends to a distance of 3 metres from a pump or a dispenser or from an opening to a storage tank at a height of 1.25 metres above the forecourt level or at the top of the pump housing whichever is the greater. It then reduces uniformly to reach a forecourt level of 4.25 metres from the pump/dispenser/storage tank opening. Figure 42 illustrates the hazardous areas and the important measurements.

Kiosks and similar buildings having openings within a Zone 2 area should be classified as Zone 2 up to a height of 1.25 metres above the forecourt level. Any electrical apparatus which is below 250 mm above the floor of the building shall be of the type suitable for installation in a Zone 1 area. Figure 42 also shows this latter zone as an area within 1.5 metres of the vent discharge pipe but the area down to ground level is regarded as a Zone 2 area for a radius of 1.5 metres from a point directly below the discharge point. Zone 1 area is also seen covering the pump housing and up to 250 mm above the forecourt level and within 4.25 metres of any pump or tank opening.

In practice, the most commonly used methods of protection for electrical apparatus are:

Zone 1   (a)   Certified Flameproof (Ex'd')
         (b)   Certified Intrinsically Safe (Ex'ib')
         (c)   Certified Increased Safety (Ex'e')

Zone 2   (d)   Any apparatus conforming to the above (a), (b) and (c), and
         (e)   Certified Type N (Ex'n')

Alternatively, the hazardous areas, notably the pump/dispenser housing, shall be certified by BASEEFA as complying with Standard SFA 3002: 1971 and labelled accordingly.

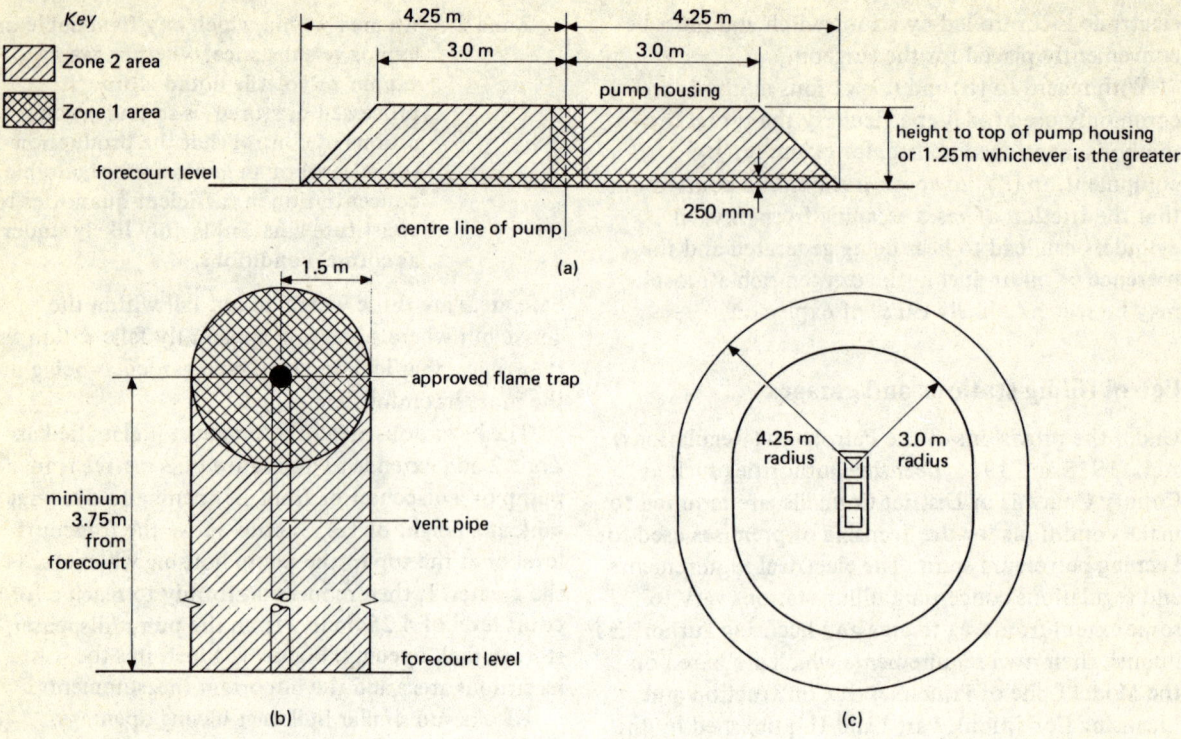

Figure 42   *Classification of the danger areas at the*
forecourt of a petrol filling station
(a) Pump housing
(b) Vent pipe
(c) Plan view from pumps

The minimum requirements within the hazardous areas on forecourts issued by a typical licensing authority are given below:

1   No provision for any socket outlets or flexible connections in a Zone 1 area unless the sockets and plugs are of flame proof design and the flexible cable is protected with a flexible metallic screen, sheathed overall. Socket outlets specifically designed for use in a Zone 2 area shall be used only if they have interlocking plugs and sockets with mercury-in-glass switches or enclosed make-break micro-switches. Alternatively, socket outlets of flameproof design may be used. No socket outlets are permitted inside a kiosk.

2   Portable and transportable apparatus, excluding handlamps, supplied by flexible cables shall be provided with earth-leakage or earth monitoring protection or both.

3   Portable handlamps for use in Zone 1 and 2 areas should be of the certified flameproof type or intrinsically safe type approved by HM Chief Inspector of Factories.

4   In kiosks, lighting fittings, switches, telephones and electrically-operated cash registers must be fitted over 1.25 metres from forecourt level and cash registers are to be permanently wired into the circuit by means of a double-pole switch with a separate flex outlet plate. Lighting fittings should be of the totally enclosed design type and heating within the kiosk shall be by means of

liquid filled radiators fitted with flameproof thermostats or other types approved by the licensing authority. They should not be installed less than 300 mm from the floor level.

5 Lighting fittings and other electrical equipment installed within the danger area associated with vent pipes, openings to storage tanks and tanker stands shall be in accordance with the specified zone classification.

6 The filling area of a forecourt, that is, an area within 4.25 metres of a petrol filling pump, shall be provided with a general illuminance of 100 lux when measured at the top of each pump or 1.25 metres from the forecourt level whichever is the greater.

7 An isolating switch shall be provided for attended self-service stations controlled from a central control point(s) to isolate the supply to all pumps. Such switches shall be additional to the switches for individual pumps. Also, where supervision of the filling station is carried out by one attendant from a central point, a loudspeaker system should be installed to enable the attendant to instruct a customer without having to leave the control point. See Figure 43.

8 Isolation is also required for unattended self-service stations, for the petrol pumps and pump lighting. It should only be capable of being restored by operation of a master switch of the contactor type and suitable for remote operation from one or more push buttons on the forecourt.

Note: The emergency switch to isolate the supply to all pumps and integral pump lighting shall be positioned as to be readily visible to the public and within easy reach for quick operation in cases of emergency (2 metres above ground level) and normally outside Zone 1 and 2 areas. It shall be marked 'Petrol Pumps – Switch Off Here' in red block letters at least 50 mm in height on a white background.

9 Where neon and/or high voltage signs are installed, a separate master switch or 'Fireman's Switch' shall be provided and suitably indicated. The signs, associated wiring, switches and other equipment shall be positioned and installed to the satisfaction of the licensing authority.

10 Wiring within the pump/dispenser housing which is not forming part of a certified intrinsically safe

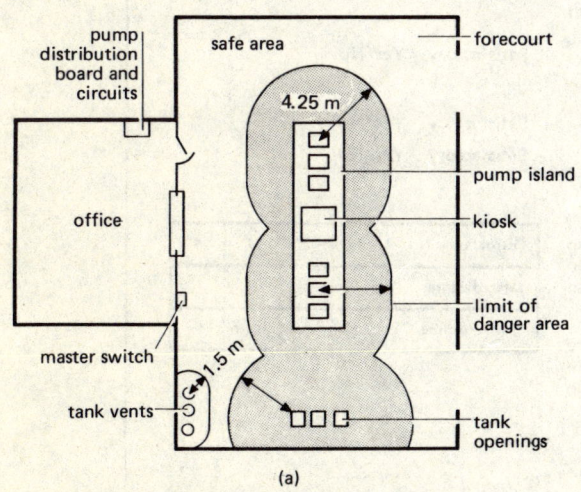 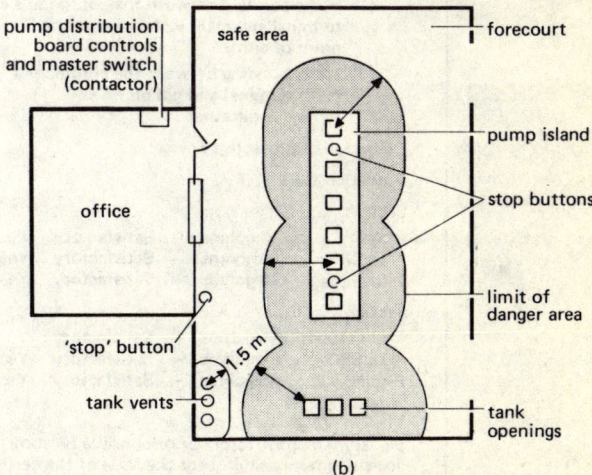

Figure 43   *Types of petrol filling station*
   (a) Attended
   (b) Unattended

XYZ COUNTY COUNCIL

PETROLEUM (REGULATIONS) ACTS 1928 AND 1936

ELECTRICAL CERTIFICATE

Name of licensee: ....

Address of licensed premises: ....

....

Date of inspection: ....   Licence no: ....

Electrical contractor's name and address: ....

....

....

I/We hereby certify that the electrical installation at the above premises appertaining to the complete petroleum installation, has been inspected and tested by me/us and complies in all respect with the requirements of the 'Model Code of Principles of Construction and Licensing Conditions (Part 1)' issued by the Home Office or the 'Code of Requirements' issued by the Association for Petroleum Acts Administration. The inspection and test includes all electrical installations within zone 1 and zone 2 areas involving petrol pumps and associated equipment, kiosks and other buildings including any place where petroleum spirit is stored, and installations within the vicinity of vent discharge pipes and delivery areas (where applicable).

The installation outside of the zone 1 and 2 areas but forming part of the petrol pump installation complies fully with the 'Regulations for Electrical Installations' issued by the Institution of Electrical Engineers, current edition.

Results of tests
(in accordance with Part 6 of IEE Regulations)

Type of earthing ....   Type(s) of protective device ....

Earth fault loop impedance..............ohms — Satisfactory   Yes/No

Continuity of protective conductors to: ·

   (1) Each petrol pump   — Satisfactory   Yes/No

   (2) All other earthed metalwork   — Satisfactory   Yes/No

Bonding:

   (1) Is the bonding between the consumer's earth
      terminal and main water pipe at the
      point of entry   — Satisfactory   Yes/No

   (2) Is the bonding between the consumer's
      earth terminal and gas pipe
      (where applicable)   — Satisfactory   Yes/No

Operation of protective device: — Satisfactory . Yes/No

Insulation tests:

Section ....

P to CPC ..... Megohms — Satisfactory   Yes/No
N to CPC .... Megohms — Satisfactory   Yes/No
P to N ...... Megohms — Satisfactory   Yes/No

Section ....

P to CPC .... Megohms — Satisfactory   Yes/No
N to CPC.... Megohms — Satisfactory   Yes/No
P to N ...... Megohms — Satisfactory   Yes/No

Petrol Pump No.	1	2	3	4
Manufacturer				
Serial Number				

Signature of contractor (or responsible person)
accepting responsibility for the issue of the certificate ....

Qualification: The appropriate electricity board or
              an approved electrical contractors on the roll of the
              National Inspection Council for Electrical Installation Contracting.

                       (delete as necessary)

Figure 44 *Electrical certificate*

circuit shall comprise one or more of the following types and generally comply with BS 5345: Part 1: 1976:

(a) Single or multicore insulated cables encased in solid drawn or seam-welded, heavy gauge screwed steel conduit, protected against corrosion, screwed into apparatus having approved glands and enclosures with all conduit connections having a minimum of five full threads engaged and a minimum axial length of thread engagement of 8 mm. Where a conduit passes from a hazardous area to a non-hazardous area it has to be provided with an approved flameproof sealing box or stopper box at the point where the conduit enters the safe area. All conduit joints should be treated against the ingress of water.

(b) Single or multicore mineral insulated metal sheathed cables terminated into accessories with approved flameproof glands. Where the cable is laid underground it shall be of the type with overall PVC sheath. It shall be laid at a depth of not less than 0.5 metres or be otherwise protected.

(c) Any other wiring system that is in accordance with BS 5345: Part 1: Section 3:25 which includes PVC armoured cables.

Note: Any duct enclosing wiring below ground shall be adequately sealed at each end so as to prevent the passage of vapour.

11 The metal enclosures of all electrical equipment, conduit runs and metal sheathing of cables shall be efficiently earthed in accordance with the current edition of the IEE Wiring Regulations for Electrical Installations.

Note: This also applies to bonding of other services.

12 Pump circuits will be dealt with as follows:
(i) Single pump or blender pump – The pump shall be provided with a single circuit to control the pump motors and integral lighting. The circuit shall be protected by a fuse or circuit breaker of suitable current carrying capacity and shall include a double-pole switch (or linked single-pole and neutral circuit-breaker) to give

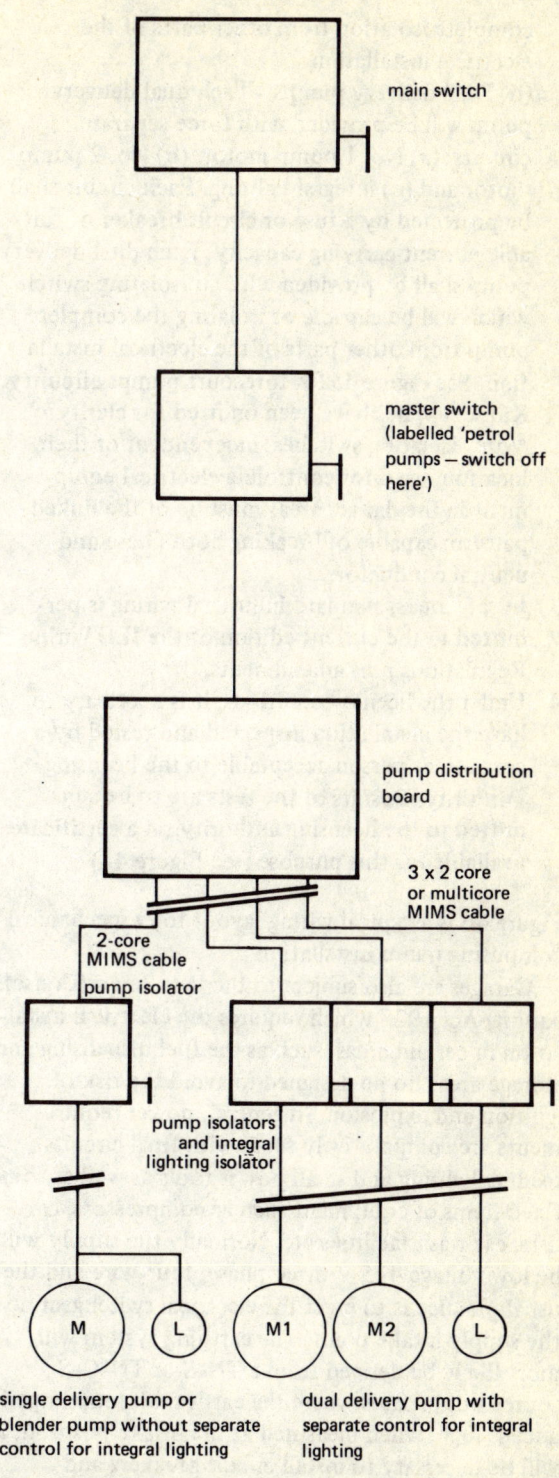

Figure 45 *Control and distribution of forecourt pumps*

complete isolation from other parts of the electrical installation.

(ii) Dual delivery pumps – Each dual delivery pump will be provided with three separate circuits; (a) No. 1 pump motor, (b) No. 2 pump motor and (c) integral lighting. Each circuit shall be protected by a fuse or circuit breaker of suitable current-carrying capacity. Each dual delivery pump shall be provided with an isolating switch which will be capable of isolating the complete pump from other parts of the electrical installation. See Figure 45 for forecourt pumps circuitry. Kiosk supplies have been omitted for clarity. *Note:* Isolating switches, independent of their location, used for controling electrical equipment in the danger areas must be of the linked pattern, capable of breaking both phase and neutral conductors.

13 In safe areas, standard industrial wiring is permitted to the current edition of the IEE Wiring Regulations plus amendments.

14 Under the license conditions, it is necessary to have the installation inspected and tested by a competent person acceptable to the licensing authority. Results of the tests are to be submitted to the licensing authority on a certificate available for this purpose (see Figure 44).

Figure 46 is a typical wiring layout for a mechanical computing pump installation.

Garages are also subject to the Petroleum (Consolidation) Act 1928 which requires the electrical installation in certain areas, such as the fuel dispensing and storage areas, to be designed to avoid the risk of ignition and explosion. In general, power requirements are comparatively small with final circuits feeding lighting and small power tools as well as some fixed items of equipment such as compressors, car lifts, car wash facilities etc. Normally the supply will be low voltage 415 V three-phase, four-wire and the usual practice is to erect the electrical switchgear near the supply intake point. The earthing system will most likely be derived from a TN–S or TN–C–S source but in cases where the earth fault loop impedance is high, when measured at the intake position, it will be necessary to install circuit breakers and residual current devices to provide better protection. The excess current circuit breaker will need to be

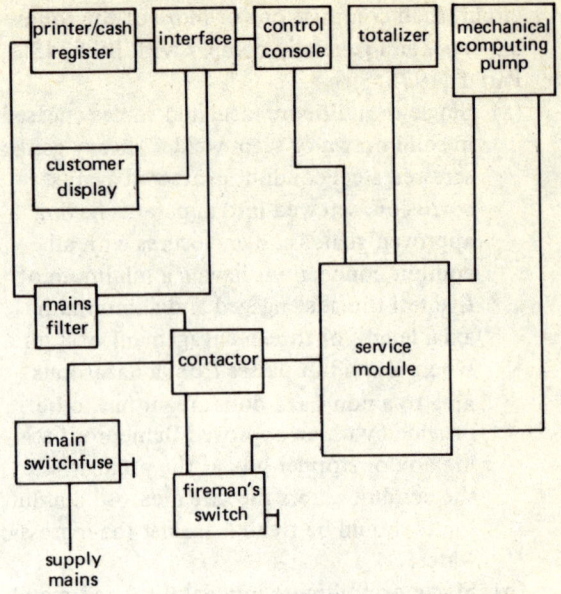

Figure 46   *Typical mechanical computing pump installation with control console*

selected to meet the prospective fault level at the intake point and the residual current devices should be designed to operate around 300 mA in 5 seconds. Where outgoing final circuits are also controlled by circuit breakers, they should have similar fault ratings to the main circuit breaker or otherwise be fitted with time delays such that the main circuit breaker would operate first in the event of a short circuit. Where fuses are fitted instead, they should be of the high breaking capacity type and also be of adequate fault rating.

From the point of view of wiring, supplies may be required at various points on the premises. This may involve several distribution boards including one for circuits operating services during non-working hours. With the exception of the offices and showrooms where PVC insulated and sheathed cables are used, most other areas will be wired using PVC cables in conduit or trunking. An alternative to this is armoured PVC cables or mineral insulated metal sheath cables having an overall PVC sheath. The latter wiring system has several advantages over other wiring systems, one of these being its heat resistance property. However, when it is used with certain equipment which produce high transients, it becomes necessary to fit surge arrestors.

With regard to lighting, some of the recommended illuminance values for different areas are: 50 lux for car parking; 300 lux for forecourts and repair shops; 500 lux for offices, showrooms and body shops, and between 750 and 1000 lux for a paint shop. There will of course be a need for supplementary lighting such as handlamps and vehicle lift lights. Luminaires installed in a paint shop should be explosion protected, that is, flame proof type, since most paints used are of a flammable nature. This also applies to the wiring system up to the *maximum spraying height*. Luminaires which are mounted not less than 2 metres above this height may be of the ordinary totally enclosed industrial type. Where purpose-made spray booths exist, the internal compartments, spraying and drying chambers, are also regarded as flammable zones and any electrical equipment used in these chambers should be of the flame proof design (type 'd' protection). In paint stores, because there is a likelihood of spillage or leak from tins, the wiring should be to Zone 2 standards. Luminaires of type 'N' protection are sufficient with control switches placed outside in the safe area. In vehicle repair shops where inspection pits exist, any fixed or portable lighting used in the pit should be of flame proof design or methods of protection suitable for Zone 1 hazardous areas.* Any handlamps used should be of the explosion protected type and capable of withstanding a 2 metre drop test – non-explosion, safety extra-low voltage handlamps are not suitable.

In terms of socket-outlet wiring, the choice is between those of BS 196 and those of BS 4343. Reference for using these can be found in Appendix 5 of the IEE Wiring Regulations. It will be noticed that BS 196 socket outlets can be used for both ring and radial circuits whereas BS 4343 socket outlets are limited to radial circuits only. It is recommended that 110 V be used to increase the safety factor but where welding sets are required, 415 V three-phase or two-phase socket outlets will be needed. Where a socket outlet is used for a car wash facility or anywhere out of doors it should incorporate a residual current

circuit breaker capable of tripping out at about 30 mA in under 30 ms. Figure 47 shows a typical method of providing reduced voltage to supply 110 V socket outlets complying with BS 4343.

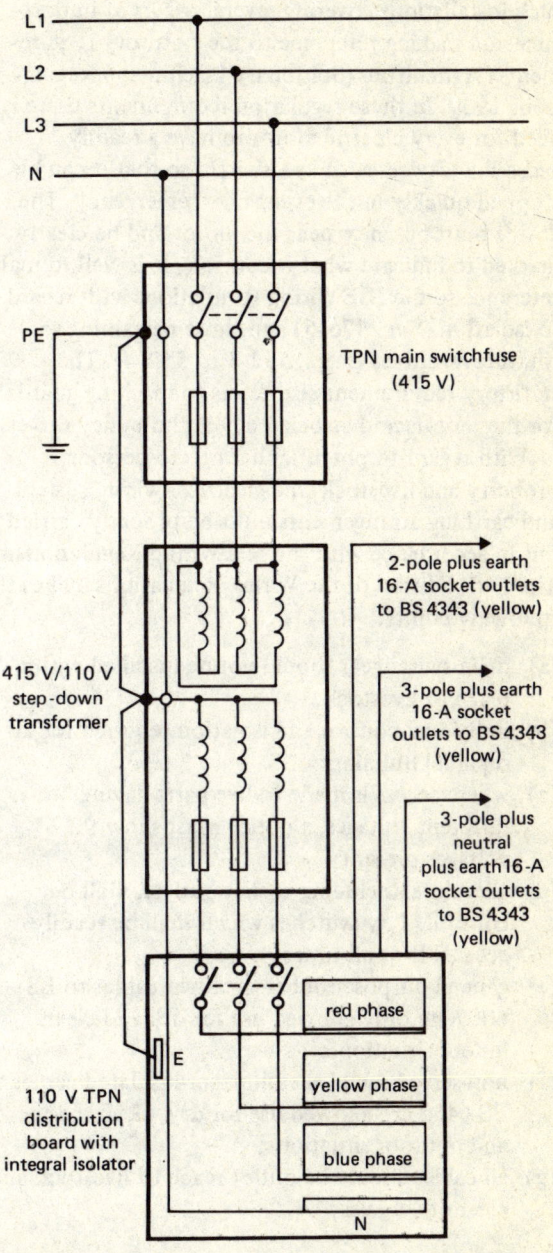

Figure 47 *Method of supplying BS 4343 110 V socket outlets*

*It is desirable to install luminaires inside the pit walls and to have no local switching. Also, the use of compressed air tools is recommended.

Agricultural and horticultural installations

Electrical installations in these premises are often regarded as potentially dangerous in view of the damp, wet and corrosive environments which exist. The scope of the IEE Wiring Regulations includes such installations covering several topics of import-ance and making reference to the statutory require-ments: Agriculture (Stationary Machinery) Regula-tions 1959. In these regulation requirements there is a need for every electric motor to have a readily accessible device, such as a switch, so that it can be stopped quickly in the event of an emergency. The device must be on or near the motor and be clearly marked to indicate what it controls. It is well to make reference to the IEE Wiring Regulations with regard to isolation (Reg. 476-5) and those pertaining to rotating machines (Regs. 552-1 to 552-4). The statutory requirement also refers to machine guards for the motors and associated belt and pulley drives.

With regard to potential hazards to persons, property and livestock, the electrical wiring system and earthing arrangements must be properly carried out in accordance with the IEE Wiring Regulations. The 14th Edition of the Wiring Regulations makes the following points:

(a) main switchgear should not be installed within reach of livestock;

(b) individual control and isolation required for all separate buildings;

(c) where access is made to live parts having different voltages, a notice must show the voltages present;

(d) all points, including socket outlets, shall be controlled by switches which shall be readily accessible at all times;

(e) general-purpose rubber sheathed cables to BS 6007 are only allowed use for dry and clean indoor situations;

(f) non-served metal-sheath paper-insulated cables to BS 6480 are allowed use for dry, clean indoor and outdoor situations;

(g) all cables should be out of reach of livestock and clear of vehicles;*

(h) cables sheathed with PVC, PCP or HOFR should not come into contact with liquid creosote;

(i) additional protection of cables may be provided by the use of non-metallic materials;

(j) external cable runs on buildings should be installed at the highest level possible;

(k) underground cables should be buried to a depth of at least 0.6 m in ground liable to cultivation;

(l) steel conduit or pipe shall not be used for spanning gaps between buildings;

(m) apparatus to conform with BS 3807 where flam-mable dusts concentrate;

(n) in garages, fixed apparatus should be installed at sufficient height/position to be clear of all vehicles;

(o) Edison-screw lampholders to be all-insulated heat-proof and drip-proof, and be provided with a protective shield;

(p) cable couplers are not permitted;

(q) switches and other control devices to be out of reach of a person in contact with wash-troughs, sterilizers etc.;

(r) in situations accessible to livestock the wiring installation should be of the 'all-insulated' construction;

(s) protective conductors are required at each outlet point;

(t) metal conduit not to be used as sole protective conductor;

(u) earthing conductor to be protected from damage by livestock or the passing of mechanical implements;

(v) isolation of metalwork is not permitted unless the metalwork is out of reach of livestock and not liable to accidental contact with passing vehicles and machinery.

Some of these requirements are illustrated in Figure 48.

Regulation 471-40 of the 15th Edition of the Wiring Regulations should be read in connection with Class II electrical equipment and where protection against indirect contact is provided by automatic dis-connection. It is worth mentioning the note pertain-ing to this regulation concerning the very low body resistance of farm animals making them susceptible to electric shock at voltages less than 25 V r.m.s. In the following regulation, Reg. 471-41, the point is made

---

*Cables liable to attack by vermin shall be of a suitable type or be suitably protected – see 15th Edition of Wiring Regulations, Reg. 523-34.

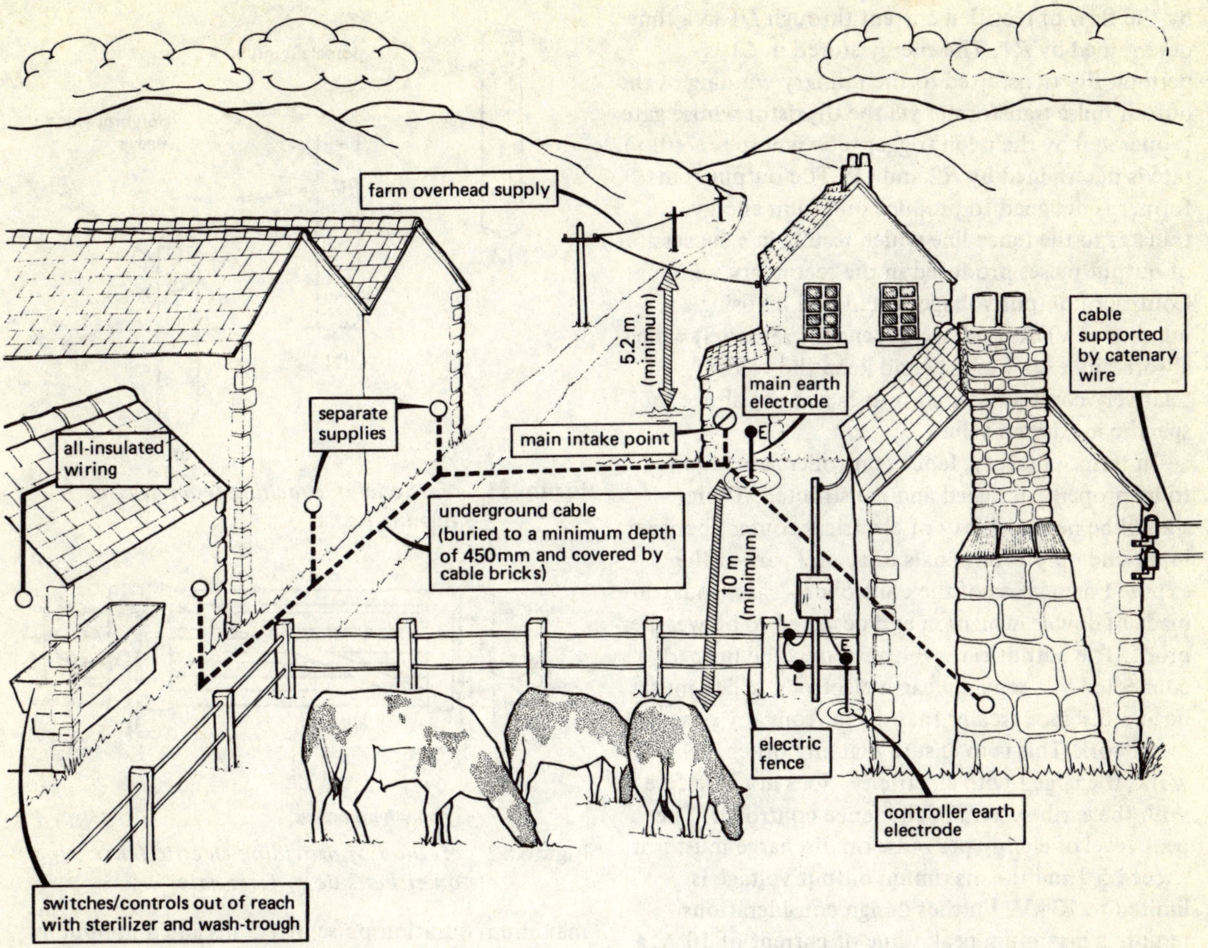

farm overhead supply
5.2 m (minimum)
cable supported by catenary wire
main earth electrode
separate supplies
main intake point
all-insulated wiring
E
underground cable (buried to a minimum depth of 450mm and covered by cable bricks)
10 m (minimum)
L
L
E
electric fence
controller earth electrode
switches/controls out of reach with sterilizer and wash-trough

Figure 48 *Farm wiring*

about using extra-low voltage in situations accessible to livestock, remembering of course that e.l.v. has an upper limit of 50 V and should be reduced to a safe value.

Where electric fences are used to contain farm animals, the fence controllers have to meet with the specifications of BS 2632 and BS 1222 (Revised). Basically, there are three forms of controller used, namely, battery-operated units, mains-operated units and mains/battery-operated units. Mains-operated units are preferable where the supply is less than 5 km away, being mounted near the supply point to reduce risks. The combined units are justified where it is necessary to keep the fence 'alive' at all times but where there is no supply available then the battery-operated units are used: the batteries being accumulators which need recharging every two weeks.

A simplified diagram of a mains-operated unit is shown in Figure 49. It will be seen that *C*1 is charged

P
R1
D1
R2
neon trigger tube
output transformer
fence terminals
240 V a.c. supply
C1
C2
R3
thyristor
N

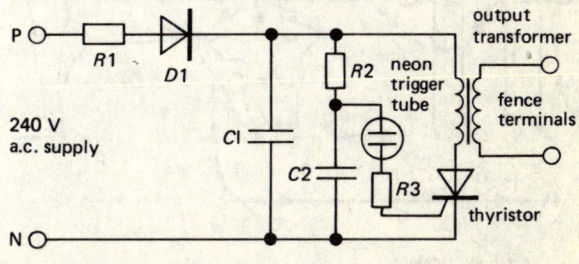

Figure 49 *Basic mains-operated fence controller circuit*

by the flow of rectified current through $D1$ in a time determined by $R1$. The energy stored in $C1$ is periodically transferred to the primary winding of the output pulse transformer via the thyristor whose gate is operated by the neon trigger tube whose repetition rate is determined by $R2$ and $C2$. The output transformer is designed to produce optimum energy transfer to the fence line which results in a succession of output pulses produced in the secondary winding. Control of output voltage is achieved by using a number of VDRs (voltage dependent resistors) which absorb any excess energy and keep the voltage relatively constant so that it does not rise above the specific maximum value.

In terms of safety, fence controllers are required to be properly designed and constructed. There should be no possibility of electrical contact between input and output terminals or to any part of the exposed metal case of the controller – cases today are made of double insulation and designed to be weatherproof. The output earth terminal must be properly connected to a separate earth electrode which must be at a distance greater than 10 m from any other metalwork. This is to ensure that any other metalwork, if charged with electricity, does not interfere with the earth system of the fence controller. The peak level of energy per pulse on discharge must not exceed 5 J and the maximum output voltage is limited to 10 kV. Further design considerations require a maximum peak value of current of 10 A, a

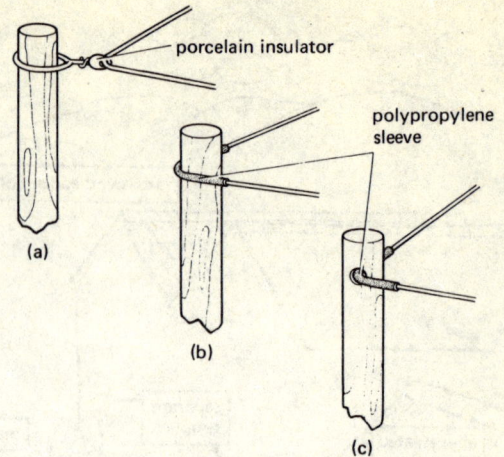

Figure 51   *Methods of attaching fence wire at turning posts*

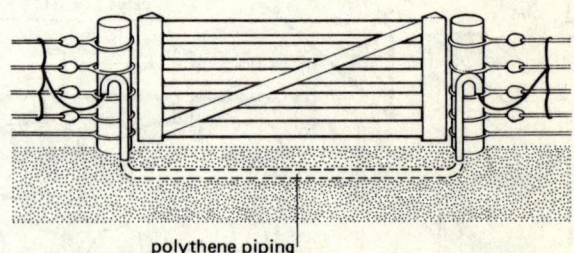

Figure 52   *Method of providing electric fence supply on either side of farm gate*

maximum duration pulse of 1.5 ms and a minimum time interval between pulses of 1 s.

With regard to the installation of both controller and fence, they are required to be installed and operated so that they cause no danger to persons, animals and surroundings. An electric fence shall not be supplied from more than one controller and barbed wire should not be used as a fence conductor since it is possible to entangle an animal in the fence where it will be subjected to continuous exposure of the electric pulses. Any fence installed along a public road has to be clearly identified with warning signs fastened to the installation at intervals not exceeding 100 m. The signs should be at least 200 mm x 100 mm saying 'electric fence' in block letters.* Figure 51 and

*The new code suggests that the warning sign shall be at least 210 mm x 105 mm in size having both sides coloured yellow and bearing a flash inside a triangle with a warning inscription outside the triangle.

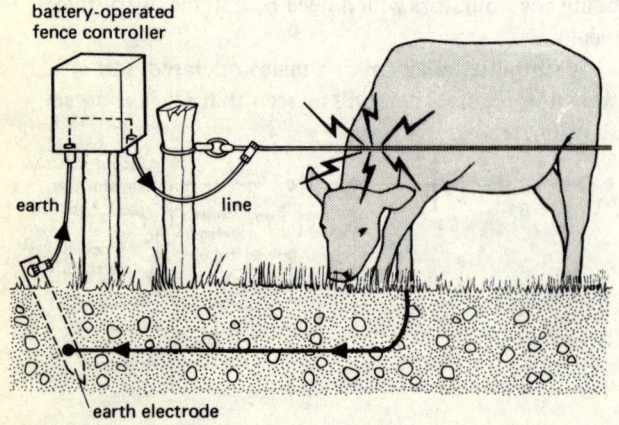

Figure 50   *Application of electrical fence*

Figure 52 show some fence installation methods. It is important not to erect a fence near an overhead power line but where this is impossible, the fence height above ground must not exceed 2 m. Also, fence wire should not be attached to telegraph poles or come in contact with metal parts not belonging to the fence installation, such as a railing. The fence wires or connecting leads must be at least 2 m from any overhead communication line. Lastly, fence and controller should be regularly inspected and maintained. Faults likely to occur are dead shorts, insulation breakdown, loose connections, deterioration of the earthing system, radio interference, faulty controllers and even a 'live' cow shed as a result of a fence line leaking badly.

Other points of interest relating to agricultural and horticultural installations are:

(a) No metalwork of wiring systems should be exposed to corrosive substances unless it is protected against such exposure. Materials likely to cause such an attack are: lime, cement, plaster and some acidic woods such as oak. In damp situations, contact between bare aluminium and any other metal having a high copper content should be avoided. All cable joints should be protected against the ingress of moisture.

(b) Cables buried direct in the ground should have an armouring or have a metal sheath protection. They should be marked by cable covers or marking tape and be buried at a sufficient depth to avoid being damaged.

Note: The 15th Edition of the Wiring Regulations does not state a minimum depth for buried cables but it recommends that this is not less than 450 mm or 600 mm where the ground is likely to be cultivated.

(c) Where an 'all-insulated' wiring method cannot be used, consideration should be given to the use of reduced voltage for portable apparatus.

(d) Earth-leakage circuit breakers or residual current circuit breakers (as they are now called) should be used where danger exists and they should be regularly tested.

(e) Where contact is unavoidable with metalwork of other services or in fact extraneous metalwork, it should be bonded to the protective conductors of the wiring system.

Fire alarm systems

Information concerning the requirements of fire alarm systems is to be found in BS 5839: Part 1: 1980: 'Fire detection and alarm systems in buildings'. The scope of this code of practice provides recommendations for the planning, installation and servicing of fire detection and alarm systems in and around buildings, ranging from simple installations to complex installations which use automatic trigger devices and sophisticated indicating equipment. The code covers systems capable of providing signals to initiate the operation of ancilliary services such as fire extinguishing systems and ventilating systems but does not cover such systems themselves. It excludes certain types of self-contained detectors, street fire alarms, 999 emergency systems as well as manually and mechanically operated sounders.

A fire alarm system may be required in a building for one or both of the following purposes, namely, (i) protection of life and (ii) protection of property. A satisfactory system is one that sounds an alarm of fire while sufficient time remains for putting out the fire or escape of the occupants before escape exit routes became unusable. Mention is made in the code of the lethal combustion products that may result from the convection movement produced by fire and which could travel to areas where people have not been alerted. Where people are actively engaged in a building, a manual fire alarm system may be sufficient, but for the protection of property it is recommended that an automatic detection system is installed as well as a manual call system in view of the fact that people could be present to raise the alarm first. The code points out that the usefulness of a fire alarm system in reducing property loss will depend also on the provision of satisfactory fire-fighting resources, for example, fire-fighting appliances and trained staff to operate them. Some further aspects of the code of practice are:

(a) circuit design
(b) zones
(c) alarm and trigger devices
(d) siting of control and indicating equipment
(e) power supplies
(f) wiring systems
(g) testing inspection and commissioning

From the point of view of circuit design, the alarm system should provide facilities for audio and visual indication when alarm signals are initiated through trigger devices such as manual call points and heat, smoke and flame detectors. Audio indication will involve alarm sounders and visual indication will involve control/indicating equipment such as indicator boards and diversion relays etc. There may also be a need to transmit signals to other equipment and other services such as the local authority fire brigade.

Triggering devices generally dictate the type of alarm system used because of their contact arrangements. They are either *normally open* or *normally*

closed. An open circuit system is one where the triggering devices are wired in parallel with each other, whereas a closed circuit system is one where the triggering devices are wired in series with each other. Figure 53 shows both these circuit arrangements. It will be noticed that the open circuit system does not need to have a line relay and it does not consume any current other than when the sounders are operating. Unfortunately, any disconnection in the line conductors will render all or part of the system out of order and this may go undetected until tests are made. Some faults such as the shorting-out of call points have the effect of making those parts of the circuit redundant; there is also the possibility of a short-circuit on the line conductors before reaching the call points. The normally closed system is to some extent self-monitoring, particularly if a break occurs in the line conductors causing the line relay to become de-energized. Both systems incorporate silencing devices for transferring the alarm signal to a supervisory buzzer when the fire is under control. The code points out that such devices should not cancel any visual indication of the alarm call while it exists.

Figure 54 shows simplified versions of monitored systems but one needs to remember that additional elements only further increase the chance of possible breakdown or circuit malfunction. In the monitored open circuit system, under normal conditions a small current passes through the circuit which is only capable of operating Relay 2 because of its high sensitivity. In doing so, the green 'healthy' lamp circuit is activated to show no abnormality. If however a disconnection occurs in the line conductors, then Relay 2 becomes de-energized and a circuit is made through the amber 'open-circuit' lamp indicating the fault. In the event of a short-circuit occurring across call points or the line conductors themselves, the Relay 1 will operate and close the red 'fire alarm' lamp circuit to signify the fault. With regard to the monitored closed circuit system, relays are often used again to detect abnormal conditions likely to occur and they will be designed having differing sensitivities in relation to the circuit elements incorporated. Generally, only one relay responds to a particular circuit condition. Under normal conditions, current flowing in the circuit is sufficient to hold Relay 1 but insufficient to operate the other two relays. A discon-

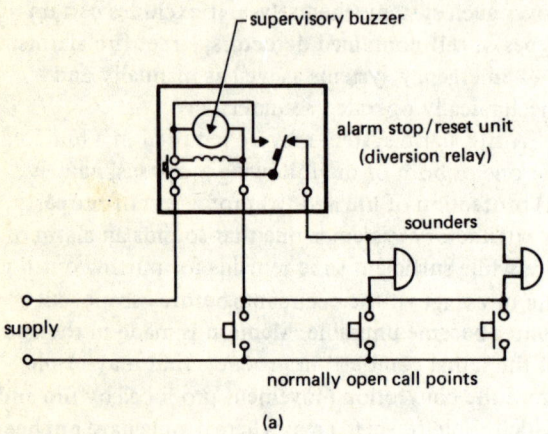

(a)

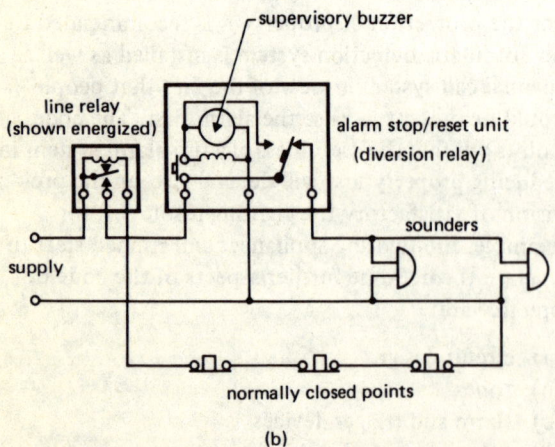

(b)

Figure 53   *Simple fire alarm system wiring*
        (a) Open circuit fire alarm system
        (b) Closed circuit fire alarm system

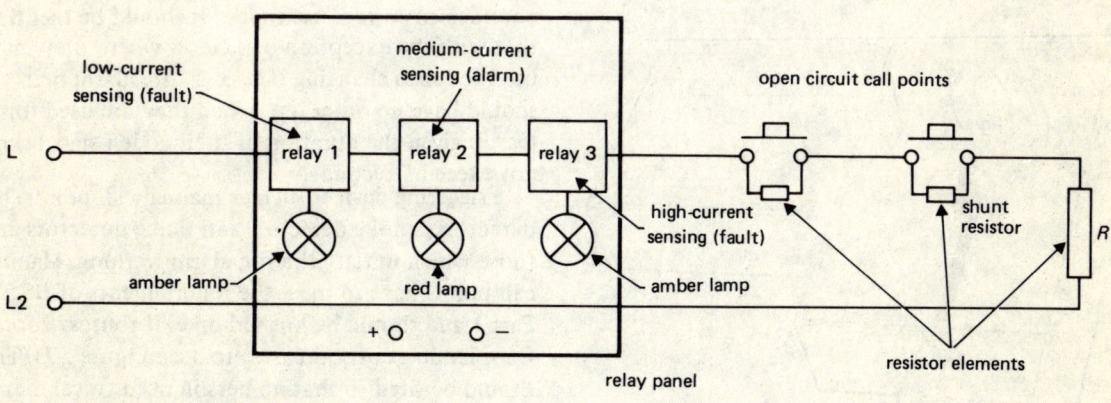

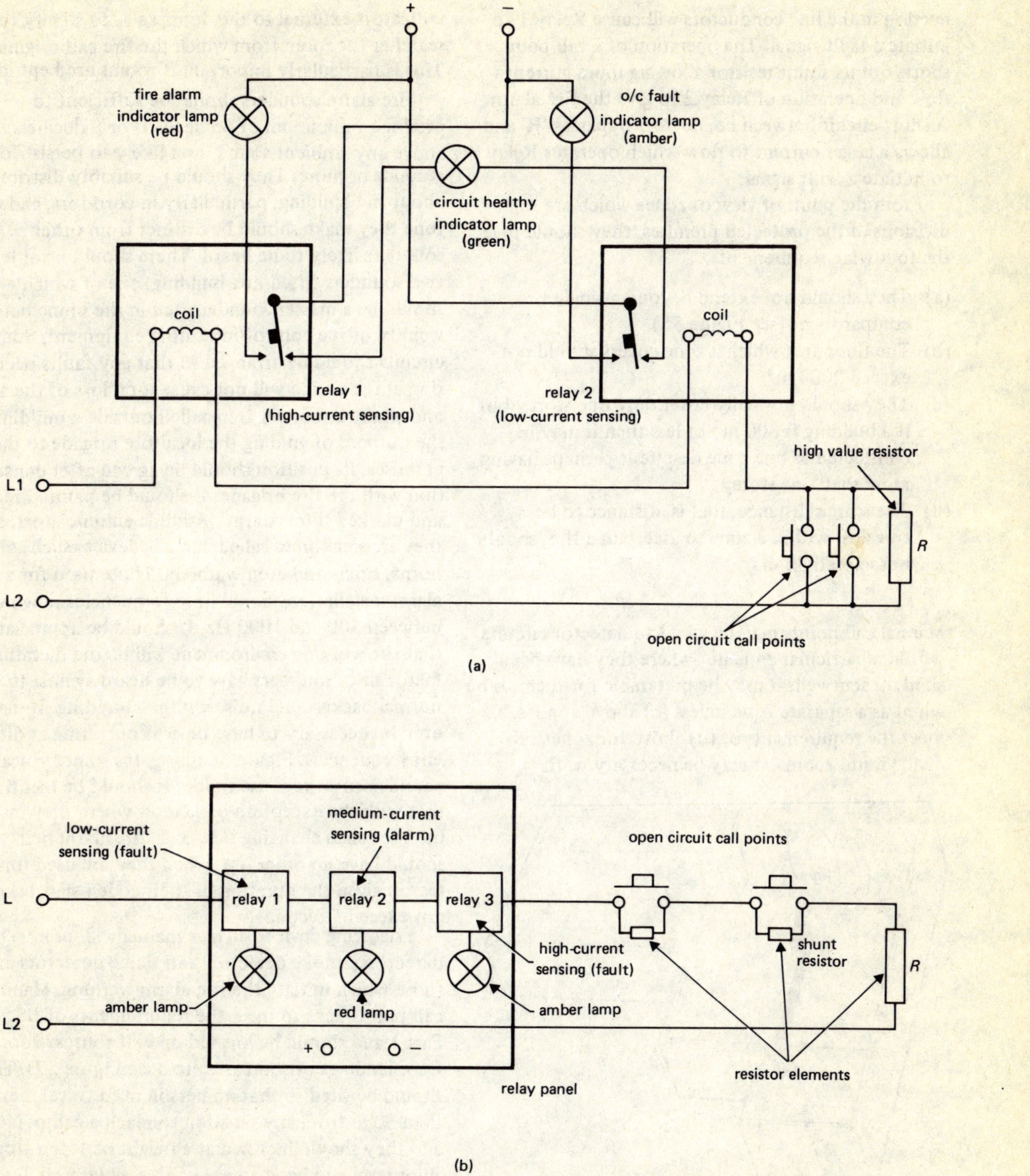

Figure 54   *Simple monitored fire alarm*
       (a) Open circuit
       (b) Closed circuit

nection in the line conductors will cause Relay 1 to initiate a fault signal. The operation of a call point shorts out its shunt resistor allowing more current to flow and operation of Relay 2 to give the fire alarm. A short-circuit between conductors bypasses 'R' and allows a larger current to flow which operates Relay 3 to initiate a fault signal.

From the point of view of zones which are sub-divisions of the protected premises, they should meet the following requirements:

(a) They should not extend beyond a single fire compartment (see Figure 55).
(b) The floor area which a zone covers should not exceed 2000 m².
(c) They should normally cover only one storey but if a building is 300 m² or less then it may be considered as one zone despite it perhaps having more than one storey.
(d) The search distance, that is, distance to be travelled within a zone to ascertain a fire, should not exceed 30 m.

Manual call points may be wired to detector circuits within a particular zone but where they have been sited on stairwells it may be preferable for them to be wired as a separate zone unless (c) above applies. To meet the requirements of (d) above for zones sub-divided into rooms, it may be necessary to fit

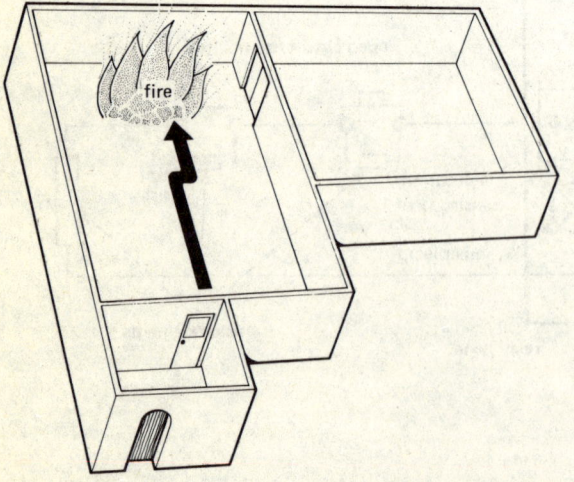

Figure 55   *Diagram of a zone*

indicators external to the doors so as to signify to a searcher the room from which the fire call originated. This is particularly important if rooms are kept locked.

Fire alarm sounders should be sufficient to produce a minimum of 65 decibels or 5 decibels above any ambient sound level likely to persist for 30 seconds or more. They should be suitably distributed about the building, particularly in corridors, and the tone they make should be distinct from other sounders likely to be heard. There should be at least two sounders fitted in a building, one of which should be a master sounder sited in the immediate vicinity of the control/indicating equipment. Sounder circuits should be arranged so that any faults occurring in the wiring will not cause total loss of the alarm and where a sounder is installed outside a building for the purpose of guiding the local fire brigade to the premises, its position should be agreed after consultation with the fire brigade: it should be painted red and marked 'Fire Alarm'. Audible annunciators, as they are sometimes called, include devices such as bells, horns, sirens and even warblers. Those used for a fire alarm installation should have frequencies ranging between 500 and 1000 Hz. It should be appreciated that the working environment will be the dictating factor since sounders have to be heard against the normal background noise within a building. It may even be necessary to have devices operating at different frequencies. Figure 56 illustrates some typical sounders in current use today. It should be mentioned that with the exception of schools where they may be used for class changing purposes, fire alarm bells should have no other use. When they are used for the reason given the duration of their coded signal should not exceed 5 seconds.

Triggering devices such as manual call points, heat detectors, smoke detectors and flame detectors are those which initiate the fire alarm warning. Manual call points have to meet the requirements of BS 5364: Part 1 and should be located on exit routes, floors, floor landings of staircases etc. (see Figure 57). They should be sited so that no person need travel more than 30 m from any position to reach a call point, and they should be fixed at a height of 1.4 m from floor level and be in an easily accessible, well illuminated and conspicuous position free from obstruction. If they are installed in a semi-flush position they must have a side profile of not less than 750 mm². Finally,

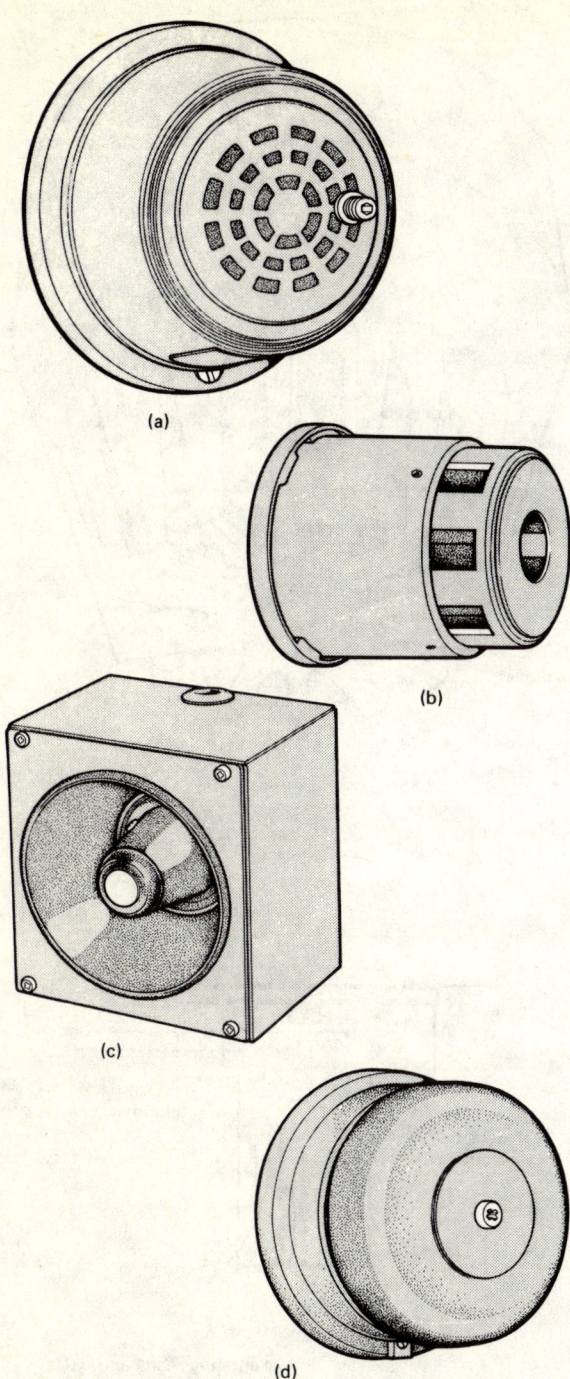

Figure 57   *Manual call point*

Figure 56   *Fire alarm sounders*
  (a) Tangent horn
  (b) Siren
  (c) Warbler
  (d) Tangent bell

when call points are operated in an automatic system, the audible warning must sound within 1 s.

In practice, there are two basic types of heat detector, one is called the *point type* the other called the *line type*. The former is affected by the hot gas layer immediately adjacent to it while the latter is sensitive to the effects produced by heated gas along any portion of the line. There are two main types of heat-sensitive element used: (i) a fixed-temperature (static) element designed to operate when it reaches a pre-selected temperature, and (ii) a rate-of-rise of temperature element designed to operate when the air temperature rises abnormally quickly. Fixed temperature types are less suitable in low ambient temperature situations or those which vary slowly. Where very high temperatures are likely to be met, heat detectors are required to comply with BS 5445: Part 8. A typical type is shown in Figure 58(b) and is set for an operating temperature of 57°C – the nominal operating temperature setting should not exceed the expected maximum ambient temperature by more than 30°C. Some conditions for its siting are: (i) heat-sensitive element is not less than 25 mm or more than 150 mm below the ceiling, (ii) should be sited more than 500 mm from walls (except where area is less than 1 m^2), (iii) maximum floor area covered should not exceed 50 m^2.

For the model shown, it is normal practice to arrange them in rows about 3 m from any adjacent wall and with a distance between rows of 6 m: this is suitable for normal ceiling heights up to 6 m. It is

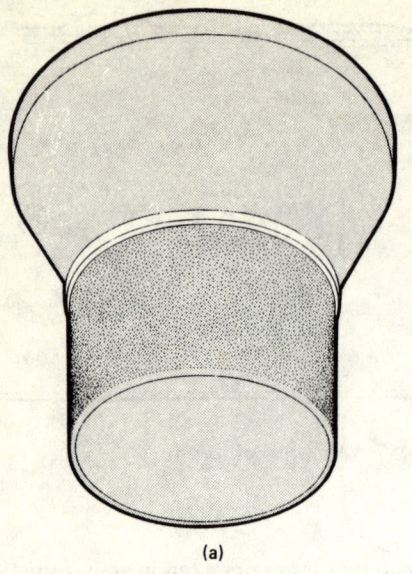

(a)

Figure 58   *Heat detectors*
           (a)   Rate-of-rise heat detector
           *   (bimetallic type)*
           (b)   Fixed temperature type to BS 5445:
           *   Part 8*

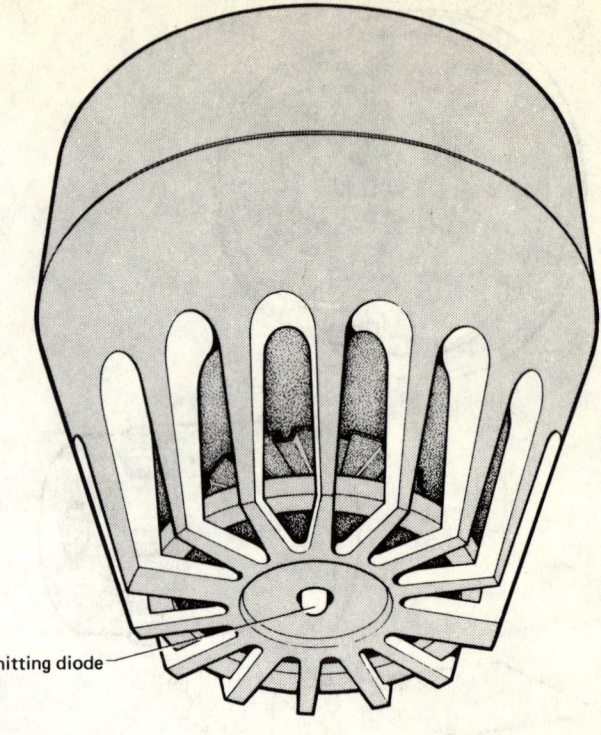

light emitting diode

(b)

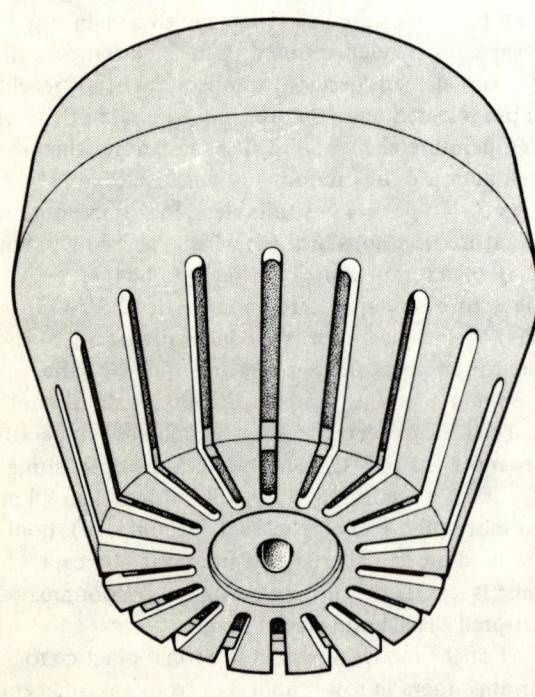

(b)

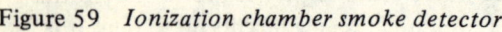

Figure 59   *Ionization chamber smoke detector*

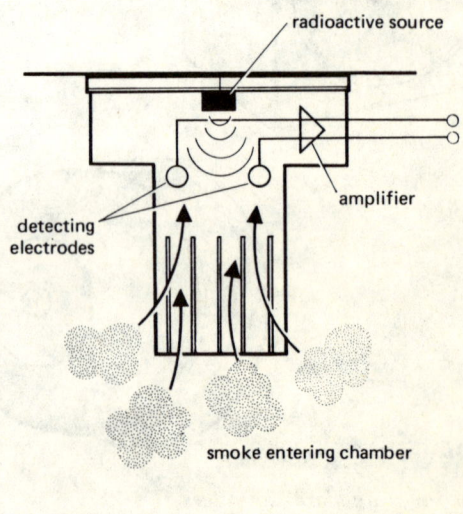

radioactive source

amplifier

detecting
electrodes

smoke entering chamber

(a)

important that no part of the detector is painted. Figure 58(a) is a diagram of another type based on the rate-of-rise principle. Its bimetallic sensing element is protected by a strong hermetically sealed metal cap. Both models are used for open circuit fire alarm systems.

There are two common types of smoke detector, namely, (i) ionization chamber type and (ii) optical smoke detector type. The first is based on current flowing through an ionization chamber. Under normal conditions ionized molecules flow freely across the chamber from the low energy radiation source to detection electrodes, thus causing a small current to flow within the ionization chamber. Ionized smoke particles move across the chamber at a lower speed and reduce the current in the chamber. When the current has dropped to a predetermined level (approximately 10 mA) the alarm circuit is triggered. Figure 59 is a typical ionization chamber smoke detector which ideally responds to smoke containing small particles produced from clean burning fires. It meets the requirements of BS 5446 Part 1: 1977. Like the heat detector in Figure 58 it has visual indication when an alarm is given; this is by means of a light emitting diode (LED) at the base of the detector grille. The optical detectors are based upon the detection of the scattering or absorption of light by smoke particles in a light beam (see Figure 60). They can be split into two groups, the *point type* and the *line type,* the former being affected by smoke in the immediate vicinity of the detector, the latter being sensitive to smoke produced on a direct line of sight to the detector.

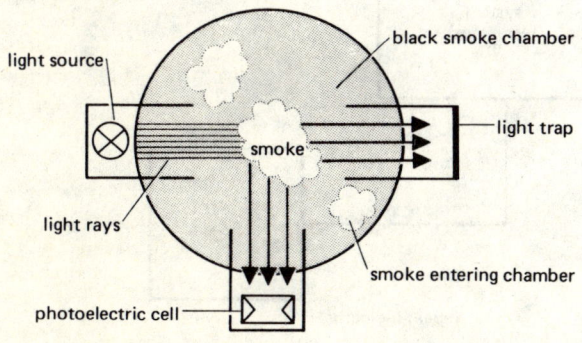

Figure 60   *Optical light scatter smoke detector*

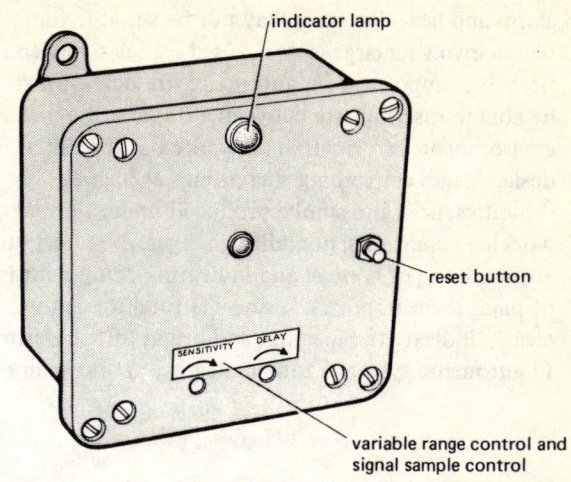

Figure 61   *Flame detector*

Automatic flame detectors detect either ultra-violet radiation or infra-red radiation. Both types employ radiation sensitive cells which 'see' the fire either directly or when radiation is reflected onto them by mirrors. Figure 61 is a typical infra-red detector which responds to infra-red radiation that has a flicker frequency between 5 and 20 Hz. Generally what happens is that the infra-red radiation is detected by a lead sulphide cell which converts the radiation into an electrical signal. This signal is then processed by electronic circuitry which monitors the output from the detector cell. If the cell produces a steady signal, say from a non-flickering source, for example, electric fire, then the detector will ignore the signal. It has to receive an infra-red signal which is flickering at a frequency characteristic of a flame, such as from a fire. When this happens an alarm relay becomes activated and operates the fire alarm system. It also illuminates an indicator lamp on the unit. This method of detection is designed for use in places such as warehouses and factories and places where intrinsically safe working is required.

It should be pointed out that the detection methods chosen should be selected on their suitability for a specific task, whether the overall purpose is for protection of life or the protection of property or both, each type mentioned above responds at a different rate and to a different fire condition. Smoke detectors, for example, provide a faster response than heat detectors but may be more liable to give a false alarm and heat detectors may not be suitable for places involving large losses caused by relatively small fires. It is important for automatic fire detectors to be able to discriminate between a fire and the normal environment. Information is required at the initial design stages concerning such things as high air velocities, heat and smoke produced under normal working conditions, humidity changes, dust, dirt etc.

The siting of control and indicating equipment is of paramount importance since its function is to clearly indicate the location and origin of the alarm. In automatic systems, zone indicators are normally displayed on a control panel often situated in an area of access such as the ground floor of a building close to the entrance point. It may be necessary to install additional indicator boards if the building is large or complex and has other entrances or night quarters and caretaker rooms. Zone indicator boards (see Figure 62) must clearly identify the signals coming into them and zones should be numbered and titled giving clear information about their location. Some control/indicating equipment may provide details of fault conditions ranging from failure of battery charging equipment to failure of scanning equipment. In

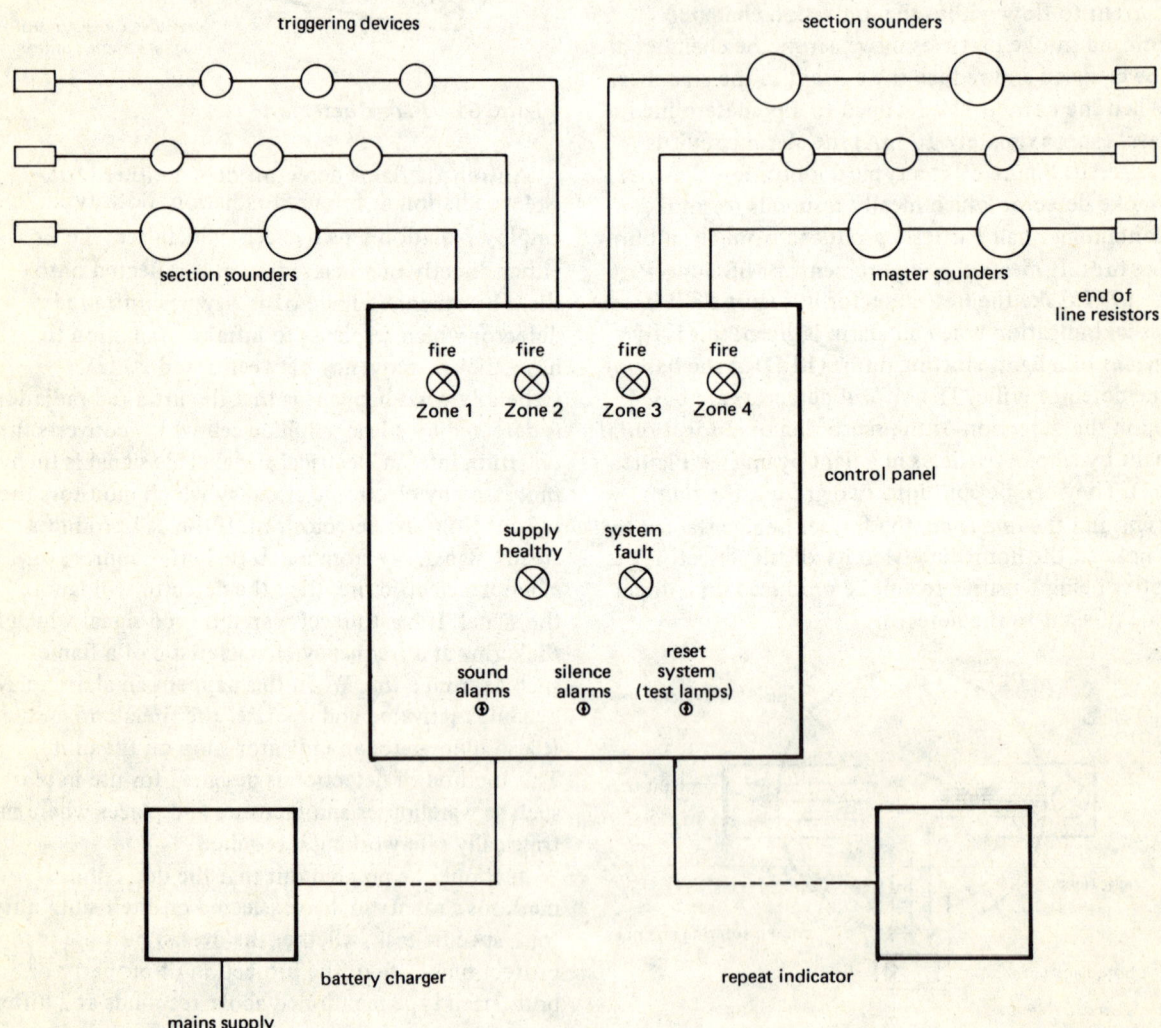

Figure 62   *Zone indicator panel*

all cases of positioning control/indicating equipment, consultation and agreement with the local fire brigade is necessary. Lastly, due consideration should be given to the ambient light levels in the vicinity of such control equipment so that it can be easily seen and operated.

The capacity of power suplies should be sufficient to supply the largest load likely to be placed on them. In practice, the most reliable supply is that of the public mains backed up by an automatic battery supply in case of failure. Such supplies should be exclusive to the fire alarm system but where they are combined with other controls or emergency systems, their reliability should not be reduced. The code also requires a switchfuse to control the fire alarm system with its cover painted red and labelled 'Fire Alarm: Do Not Switch Off'. The electricity supply to the switchfuse should be so arranged that continuity of supply is ensured at all times, even if the supply is switched off due to the building being unoccupied or for economy in consumption of power.

The code specifies the requirements for the duration of standby supplies and battery life for premises with and without standby generators. Whatever the case, the standby supplies should be capable of maintaining the system in full normal operation for at least 24 hours and at the end of that period still have sufficient capacity to provide the evacuation alarm in all zones for a further 30 minutes. Battery provision should allow for any future extension of the fire alarm system and choice between batteries having a service life of at least twenty years (Plante lead-acid and nickel alkaline types) and those with a life of about four years, depends mainly on economic factors.

With regard to wiring systems, the code distinguishes between cables likely to continue operating when attacked by fire and those not required to continue operating for any appreciable time when attacked by fire. Certain cables meet both these conditions, namely, MICC cables to BS 6207, PVC-insulated non-sheathed cables to BS 6004 and general purpose elastomer-insulated textile-braided and compound cables to BS 6007. Where conduit, ducts, channels, or trunking is used for the cables of fire alarm circuits they should be reserved exclusively for those cables. Alternatively, cables should be separated from cables of other circuits by rigid, continuous

screens of non-combustible material, but where this provision cannot be met the fire alarm system should be wired in MI cable as outlined above. Where wiring systems pass through floors and walls etc., the surrounding hole should be made good with fire-resisting material to the full thickness of the floor or wall etc., and internal fire barriers should be installed inside channels, ducts and trunkings at these points in the building.

Other points of interest concerning cables and wiring are: precautions taken where cables or metallic conduits are installed in damp or corrosive situations; use of a suitable and accessible junction box labelled 'Fire Alarm' to avoid confusion with other circuits; adequate conduit/duct/trunking sizes; adequate earthing arrangements, and compliance with the IEE Wiring Regulations. Also, due regard must be paid to the limitations imposed by voltage drop when selecting conductor sizes, particularly those systems operating at 24 V.

Like all electrical installations, the fire alarm system has to be inspected and tested on completion to ensure that the work has been carried out in a satisfactory manner. The installer is required to issue a certificate which specifies that the system complies with the code of practice and where deviations exist they must be stated. While normal insulation testing has to be carried with a 500 V d.c. tester in accordance with the Wiring Regulations, it should be carried out with all equipment disconnected to avoid damaging electronic components. Where a system has been completed, voltages in excess of the normal operating voltage cannot be used but the earth insulation resistance can be checked with a voltmeter connected between earth and each pole of the battery supply; any deflections given will indicate a faulty insulation resistance. Besides inspection and testing, the fire alarm system requires commissioning tests checking such things as triggering devices, operation of alarm sounders, control/indicating equipment and ancillary equipment.

The code makes recommendations for the transmission of alarms to the local authority fire brigade by the speediest and most certain means available. Some of these methods are: (i) *direct* – through private PO line; (ii) *PO operator* – through manual or autodialler 999; (iii) *direct or PO operator* – through alarms by carrier (ABC), and (iv) *central station operator* – through private PO line to central alarm station or public switched telephone network. The first method is very reliable causing no delay and a minimum of human intervention but long lines can be expensive and the system is unacceptable to many fire brigades because of the time required for testing and advising fault warnings. The method can be used feeding into VF 'System A' (voice frequency remote control and communications system) with relatively short lines but delays occur if equipment receives more than one call simultaneously. The second method of communication is designed to guard against failure to receive an emergency call. The manual 999 provides for the receipt of an alarm being confirmed by the fire brigade and therefore requires a continuously manned centre at the protected premises. The 999 autodialler uses the emergency system via the public switched telephone network and on connection plays a recorded message. The

mechanical replay equipment could possibly limit its reliability. The third method uses existing PO lines so that coded signals from fire alarm equipment are sent to the local exchange and fed into a multiplex system. The signals are received at a central exchange and then delivered to action points such as equipment at the fire brigade control room. The fourth method can be economical, particularly the public switched telephone network in which coded signals from the fire alarm system are sent over the network to an automatically dialled control station.

Standby supplies

Disruption in the public electricity supply has demonstrated many times over that standby power supplies are essential in certain premises. In some instances failure of supply cannot be tolerated for even a brief moment. This is often the case with electronic data

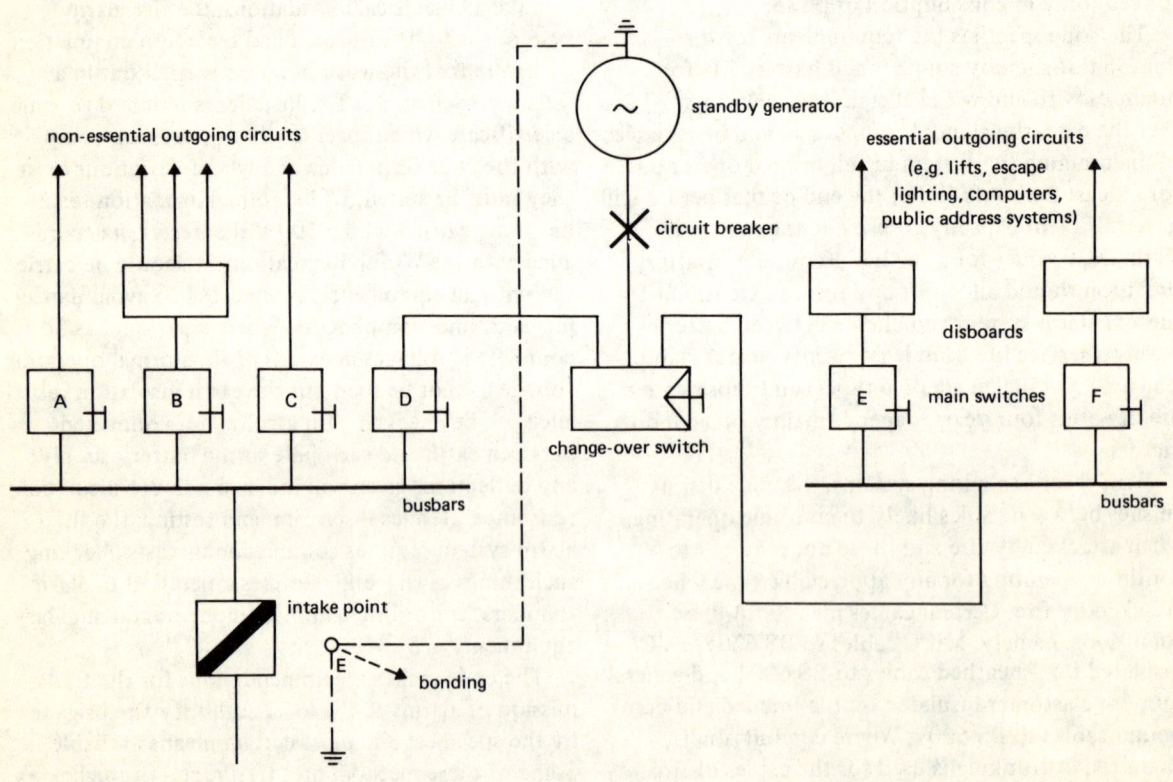

Figure 63   *Wiring arrangements for emergency services using four-pole change-over switch*

processing and computer systems, air traffic control systems, telecommunication systems, fire alarm systems, to name but a few. For large industrial users of electricity, continuity of supply is vital and stand-by supplies will enable production to proceed as well as keep important plant operating until normal supply is restored. An example of this might be the fans and burners associated with a furnace.

In practice, choice of standby supply mainly rests between battery sources and generator sources. In determining this choice a number of factors will need to be taken into consideration. Battery sources merely store a fixed quantity of electricity and are often seen as a *temporary* arrangement finding considerable use in situations where the emergency load is relatively small and the period of operation is relatively short. Generator sources, on the other hand, are normally powered by internal combustion engines and provide a continuous supply, dependent

only upon the maintenance of fuel supplies. Situations calling for their use are those demanding a much greater time extension and a much greater load requirement as well as requiring normal supply voltage and frequency.

Figure 63 shows a simplified wiring arrangement of a standby generator connected into the system through the operation of a four-pole change-over switch. The essential outgoing circuits will normally be fed from the main switch (D) and on failure of the supply the change-over switch is operated to bring in the standby generator. Modern practice is to use mechanical interlocks between the supply mains and generator (or alternator) as shown in Figure 64. In this arrangement the control circuit is designed to monitor the mains supply, to start the engine in the event of complete or partial failure and also connect the load to the alternator. This cycle usually takes about 7 to 10 seconds. When the supply is eventually

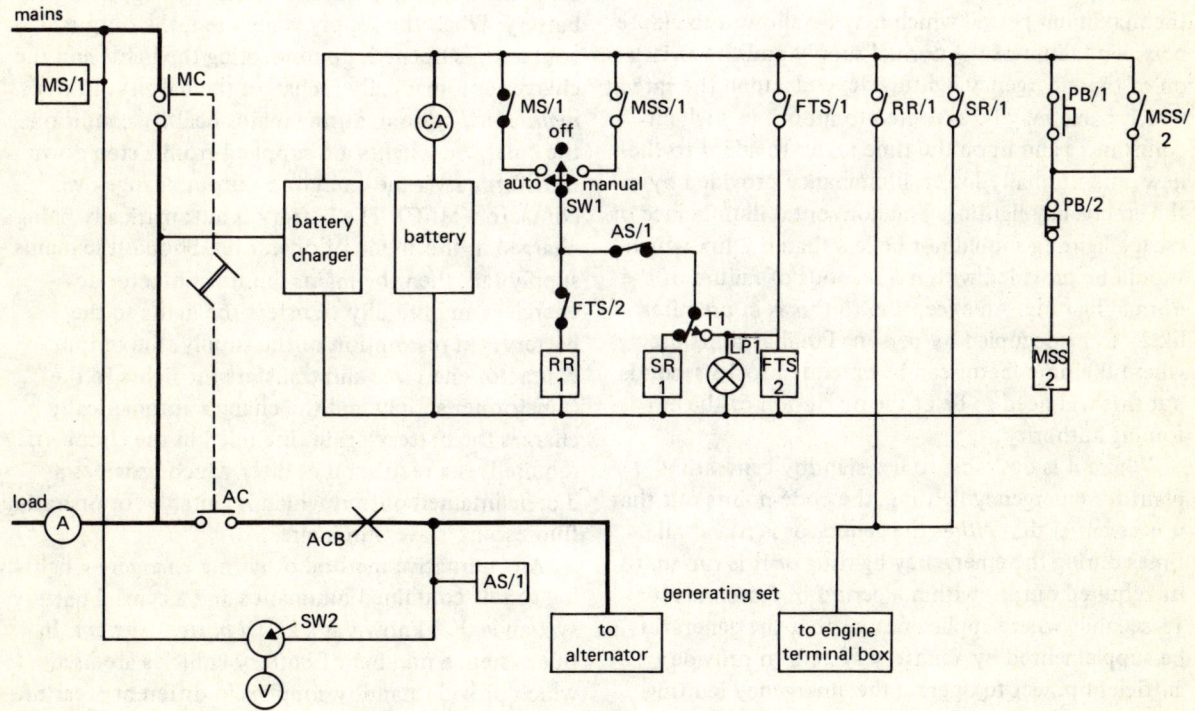

Figure 64   *Basic mains failure system showing mechanical interlock*

restored, the load is transferred back to the mains and the engine stopped. If at any time the engine fails to start, the unit will shut down after a pre-set period (approximately 10 seconds). When this happens a 'failure-to-start' lamp operates. Provision is made for the starter battery to be trickled-charged from an automatic mains failure control panel while the supply is normal. Provision is also made for the stand-by generator to be started by hand for maintenance and test purposes.

Today, there are many types of standby generator available; they range from the simple manual start/stop sets which are often portable or transportable to the large self-contained sets for permanent indoor or outdoor use.

In Volume 2 of this series, mention is made of emergency lighting and it is felt a little more could be said on the subject now that ammendments have been made to BS 5266: Part 1: 1975, that is, 'Emergency lighting' – of premises other than cinemas and certain other specified premises used for entertainment. The code of practice mentions that the determination of the maximum period which may be allowed to elapse between failure of the normal supply and the switch-on of the emergency lighting depends upon the rate at which panic may be expected to mount in such circumstances and upon the time taken to adapt to the new, and normally lower, illuminance provided by the emergency lighting. The horizontal illuminance of escape lighting should not be less than 0.2 lux which should be provided within 5 seconds of failure of the normal lighting. An exception to this is in premises likely to be occupied by persons familiar to its use where the time factor can be extended to 15 seconds but this will need to be at the discretion of the enforcing authority.

Where it is desirable to use standby generating plant for emergency lighting, the code points out that it is essential that *either* the generator is run at all times during the emergency lighting *or* it is run up to its required output within a period of 5 seconds (or 15 seconds where applicable); or that the generator be supplemented by a battery system to provide sufficient power to operate the emergency lighting system for at least 1 hour.

Where, however, the need for standby generating plant is not of prime importance, choice of emergency power is often left between self-contained luminaires and a central battery system. Self-luminous signs may be used with either system. The main advantages with using self-contained luminaires are their individual cost and their relatively speedy installation time with minimum disturbance to the existing decor – such luminaires contain their own battery, charger and control circuit. They are beneficial in small installations and those that can be easily extended by adding more luminaires. The main advantages with a central battery system are its wider range of duration and wider choice of luminaire as well as offering choice of battery to suit individual requirements. The system also provides a means of central control together with testing facilities and these factors make it more suitable for use in large premises.

Figure 65 shows two methods of wiring emergency lighting using a central battery system. In the *non-maintained* system, under mains healthy conditions, the emergency lights are off and the battery charger operates to keep the battery in a charged state. In the event of supply failure, the output contactor immediately closes, connecting the emergency lights to the battery. When the supply is restored, the output contactor is opened, disconnecting the lights and the charger automatically recharges the battery. In the *maintained* system, during mains healthy conditions, the emergency lights are supplied from a step-down transformer via the energized output change-over contactor (MFC). The battery is automatically being charged in this mode of operation. Should the mains supply fail, then the mains failure contactor de-energizes and initially transfers the lights to the battery. On restoration of the supply, the output contactor energizes and transfers the lights to the transformer supply and the charger automatically charges the battery again. Included in the circuit (if required) is a rectifier d.c. filter which provides a d.c. maintained output which is suitable for operating fluorescent 'slave' luminaires.

An alternative method of wiring emergency lighting to self-contained luminaires and a central battery system is that known as a *zonal battery system*. In this system a number of battery cubicles are used which provide standby supplies in different areas of the premises. The battery cubicles are self-contained and can be easily installed with 'slave' luminaires containing only lamps and no control gear. This has the advantage of minimizing long cable runs and helps

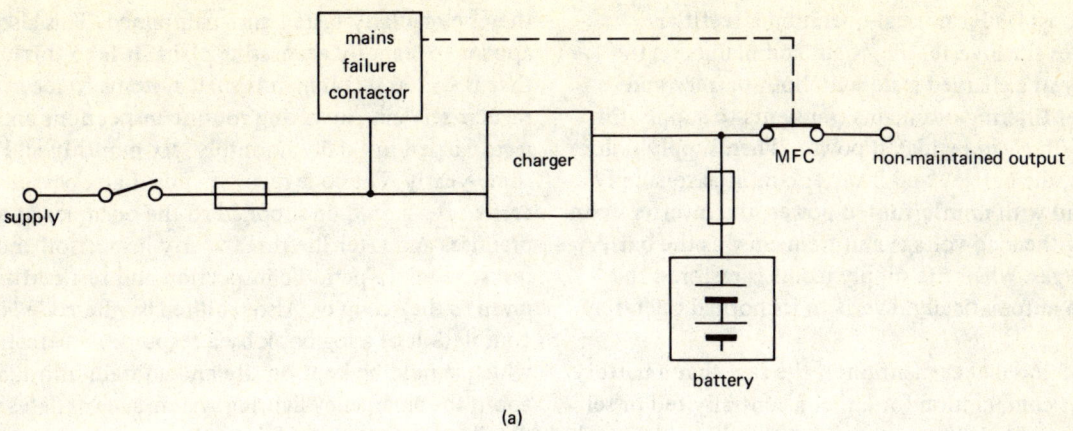

(a)

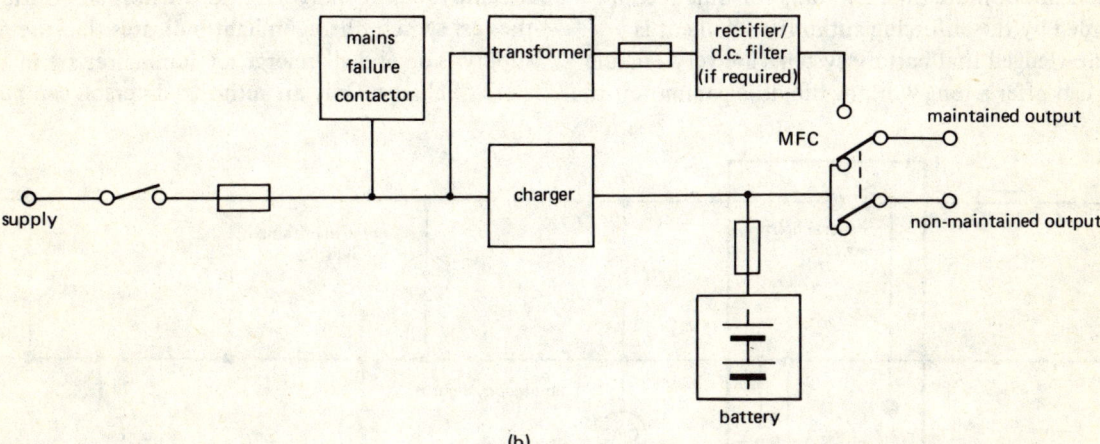

(b)

Figure 65   *Central battery systems*
   (a)   Non-maintained
   (b)   Maintained

to reduce volt drop problems.

Where the emergency lighting source is by tungsten-filament lamps, battery energy may be fed to the lamps as direct current, but where fluorescent tubes are used, the battery output can be converted to alternating current by a static inverter. Figure 66 is typical of this arrangement but is more suited for industrial control, data processing systems, telecommunication systems and security systems. It serves as an illustration of what a static inverter is and how it

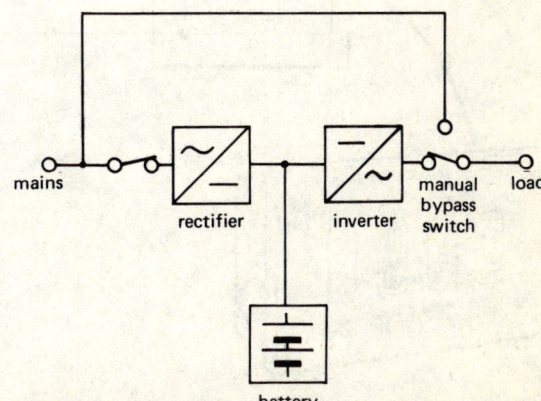

Figure 66   *No-break static inverter system*

functions. Under normal operation a rectifier provides the inverter d.c. input and maintains the battery in a charged state with both battery and inverter filtering out mains transients to supply the load with *clean* regulated power. When supply failure occurs, the battery and inverter continue to supply the load with uninterrupted power, the inverter maintaining the load voltage and frequency as the battery discharges. When the supply mains is restored the system automatically reverts to its normal operating mode.

BS 5266 makes mention of the fact that a battery/ charger combination for either a centrally fed or self-contained luminaire arrangement should be designed so that after the battery has been discharged for its specified duration (which is normally between 1 hour and 3 hours), it should be capable of again supporting the specified duriation period following a recharge period of not more than 24 hours (or time recommended by the enforcing authority). While it is acknowledged that battery systems are very reliable and can offer a long working life, it is paramount that

they be regularly tested and maintained. This also applies to standby generating plant. It is to this end that BS 5266 recommends such systems to receive regular servicing involving routine inspections and tests carried out daily, monthly, six-monthly and three-yearly. The code recommends that a completion certificate should be supplied to the occupier of a premises and after the three-yearly inspection and test schedule, a periodic inspection and test certificate given to the occupier. Also required by the code is the completion of a log book by a responsible person which should be kept on site and contain information about the emergency lighting system such as dates and details of defects and action taken as well as any alterations to the system.

Figure 67 shows a method of testing single-point (self-contained) luminaires. It ensures that the emergency lighting circuit is tested without disconnecting the general lighting. In the normal position of the test switch, the neon light indicates that the mains supply is on and all emergency luminaires are in a state of charge. Only an authorized person can put

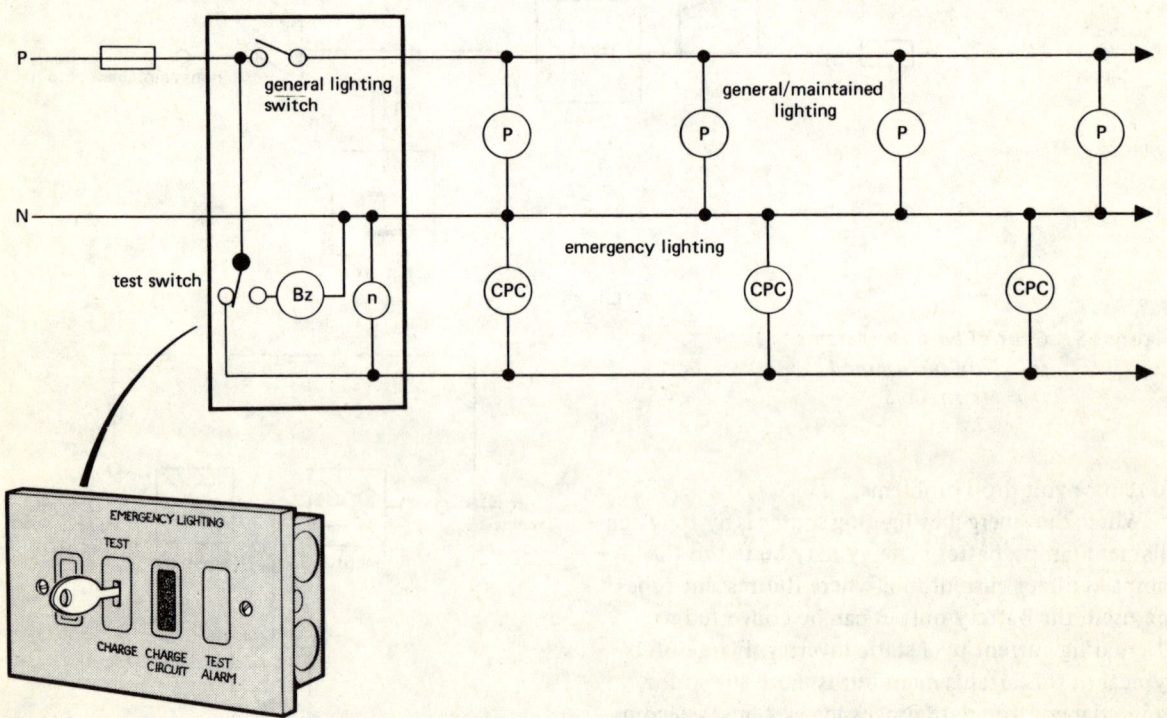

Figure 67   *Method of testing single point luminaires with test switch in normal position*

the system in the test position using a special key. In this mode the neon will not light and it is the object of the a.c. buzzer to give a permanent indication that the system is being tested. The reason for providing an audible alarm is that there may be a risk with a silent switch that somebody may leave the circuit in the test position whereupon the batteries in the emergency luminaires will discharge and not provide cover at a time of real emergency. The buzzer, then, ensures that the circuit cannot be left in the discharge position by accident.

In summary, it can be said that choice between a central battery system and single-point system will have to consider such factors as: equipment costs, installation costs, and long-term maintenance costs. Some manufacturers recommend that small buildings with less than twelve points tend to be best suited for single-point installations but this may not be the only factor since larger spread out buildings might find this method an advantage on account of the high distribution costs of a central battery system. With regard to maintenance costs, consideration has to be given to the anticipated life of the building and its installation. Sealed nickel cadmium batteries have to be regularly changed at intervals between five and seven years. In practice, batteries used in single-point systems should be replaced every five years whereas in a central battery system the life of batteries can be very long, providing the advantage of a low system cost and also battery cost per watt. With this system, however, care must be taken to ensure that there is no excessive volt drop between the battery system and luminaires. It may be advantageous to use the wide voltage range of fluorescent luminaires by increasing the voltage output of the battery system. Figure 68 shows a number of ways of providing emergency lighting for a premises.

Note: At the time of writing a further two chapters concerned with supplies for safety services have been added to the IEE Wiring Regulations (see Figure 1).

Lightning protection

Protection of structures against lightning is covered in British Standard Code of Practice 326 (1965). Here it is pointed out that a lightning discharge current could

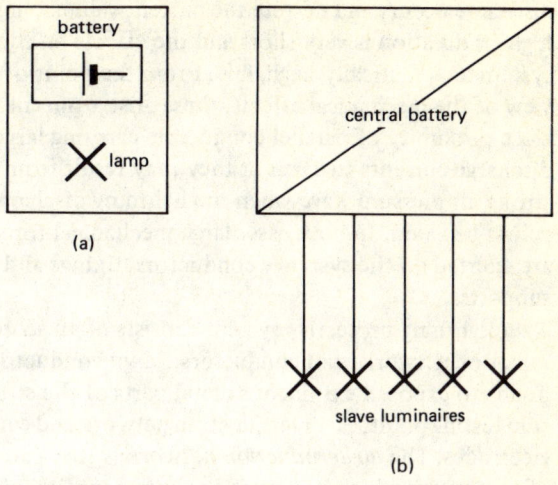

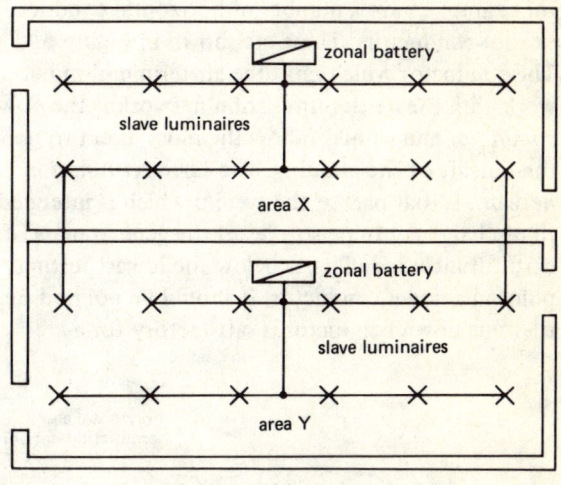

Figure 68   *Methods of providing emergency lighting*
　　　　　　(a)   Single-point luminaire
　　　　　　(b)   Central battery system
　　　　　　(c)   Zonal battery system

reach a maximum 200 000 A, producing all three electrical, thermal and mechanical effects.

The electrical effects are made manifest by the momentary raising of the lightning protective system's potential, with respect to true earth, to a high value that may possibly cause *flashover* to metalwork either on or in the structure being protected. The thermal effects are related to the temperature rise of the conductor through which the discharge occurs and despite the current value being high its duration is very short and the effects on the system are practically negligible. From the point of view of the mechanical effects, these arise from the close proximity of parallel conductors carrying large discharge currents to earth or they may result from a strong air pressure wave when the lightning discharge is first initiated. In both cases large mechanical forces are exerted on the system's conductors, fixings and supports.

A lightning protective system consists of an air termination network, roof conductors, down conductors, bonds to exposed extraneous metal parts of the structure testing point, earth termination network and earth electrodes. The *air termination network* is that part of the system which is intended to intercept lightning discharges and it may comprise a vertical conductor (VC) or horizontal conductor (HC). Large buildings, for example, have a number of horizontal conductors or roof conductors. These are shown in Figure 69. The conductor which links the air termination network with the earth termination network is the *down conductor* and should follow the most direct route on the outside of the building. The *earth termination network* is that part of the system which is intended to discharge lightning strokes to the general mass of earth: it includes all parts below the lowest testing point in a down conductor. It should be pointed out that one down conductor is satisfactory for a

structure having a base area less than 100 m² or a non-conducting chimney which is less than 1.5 m across its top. However, one additional down conductor should be provided for every additional 300 m² of the base area or every 30 m of perimeter, whichever requires the least number. Such conductors are made of either copper or aluminium strip (20 mm x 3 mm) which is often covered with PVC for technical or aesthetic reasons. Fixing saddles or clips should be spaced at intervals not exceeding 1 m so that the weight is evenly distributed over the fixings. Each down conductor should have a test joint in it and be conveniently positioned where it cannot be interfered with by unauthorized persons. The test clamp should be made of the same material as the conductor (or an alloy complying with the appropriate British Standard) and it should incorporate facilities to enable the test conductor to be readily clamped to the down conductor besides ensuring good electrical contact. Figure 70 is a typical test joint having a centre bolt test clamp.

With regard to roof conductors, these should be fixed along ridges or around the roof perimeter and interconnected with other conductors so that no part of the roof is more than 9 m from a conductor. Again, copper or aluminium conductors may be used and should be fixed as before but additionally allow sufficient clearance around the conductors for expansion and contraction. Lower buildings or structures which are within the *zone of protection* need not be separately protected. The zone of protection (ZP) is the area or space within which a lightning conductor provides protection against direct lightning strokes by directing the strokes to itself. These areas, for vertical

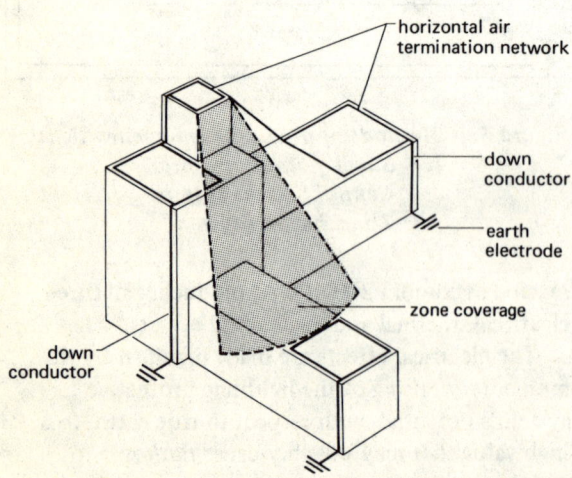

Figure 69   *Lightning protection of large buildings*

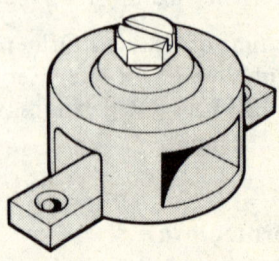

Figure 70   *Centre bolt test clamp*

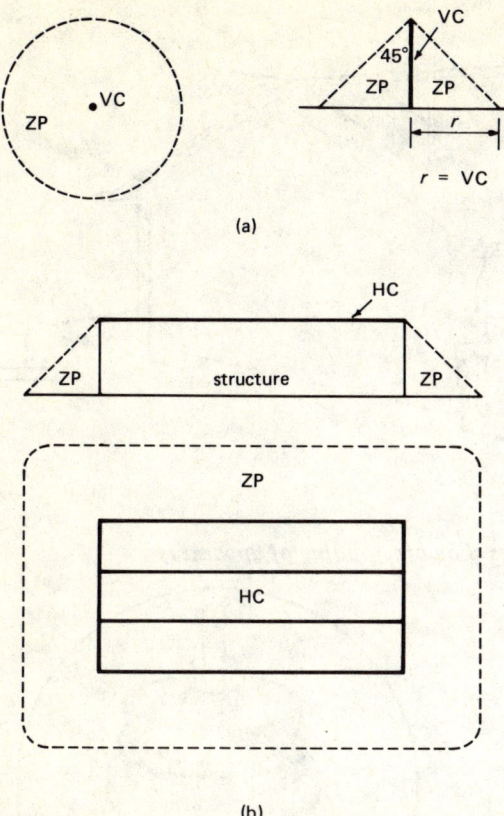

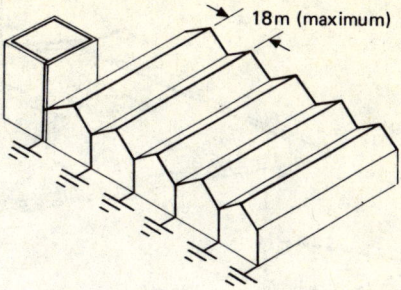

Figure 72 *Air termination network showing arrangement of parallel conductors*

Figure 71 *Zone of protection for vertical and horizontal conductors*
(a) Zone protection of a single vertical lightning conductor
(b) Zone protection of horizontal lightning conductors

and horizontal conductors, are shown in Figure 71. For practical purposes it is assumed that a lightning conductor will protect an area below it equal to a 45° cone with its apex at the highest point of the conductor. Thus, the base diameter of the cone is twice the length of the vertical conductor. The code (CP 326) points out that between two or more vertical conductors, spaced at a distance not exceeding twice their height, the equivalent protective angle may be taken as 60° to the vertical. From a design point of view, it is important to realize that a single conductor

may not fully protect a whole structure. Because of this, large roof premises are fitted with a number of parallel horizontal conductors as shown in Figure 72.

A lightning protective system will only operate efficiently if it has been installed correctly. To this end, all metallic projections such as radio and television masts, chimneys, vent pipes etc. should be bonded to parts of the air termination network (see Figure 73). Where dissimilar metals are involved there should be complete waterproof protection of the joint and effective insulation of the conductor. Where joints and bonds are made, surfaces should be thoroughly cleaned, coated with petroleum jelly or other suitable anti-corrosion compound and adequately mated with some form of high pressure clamping. Such clamping material should possess the same electrochemical potential to that of the conductor material if possible – see 'Corrosion' page 79. Figure 74 is typical of the bonding within a premises supplied from a TN–C–S source. Not only does it provide potential equalization but also overvoltage protection in the form of surge arrestors. This method is used for lightning protection of premises in potentially explosive environments.

With regard to earth resistance, the whole lightning protective system should not exceed 10 Ω. Where there is more than one down conductor, the resistance of each individual earth termination should not exceed 10 times the number of earth terminations. A reduction of a resistance to earth below 10 Ω has the advantage of reducing the potential gradient around the earth electrode when lightning discharges occur.

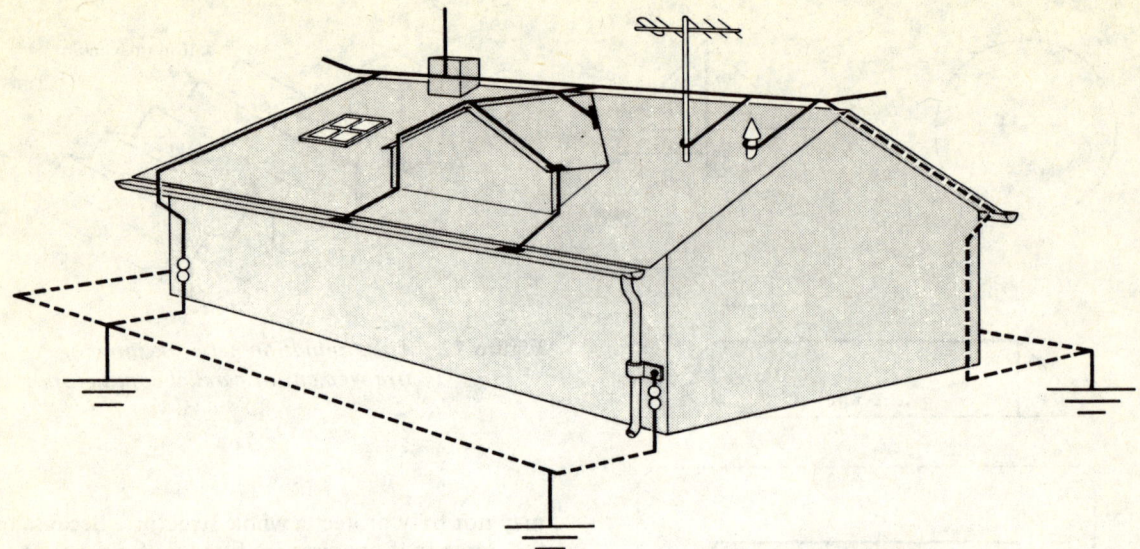

Figure 73   *External lightning protective system of a premises showing bonding of apparatus*

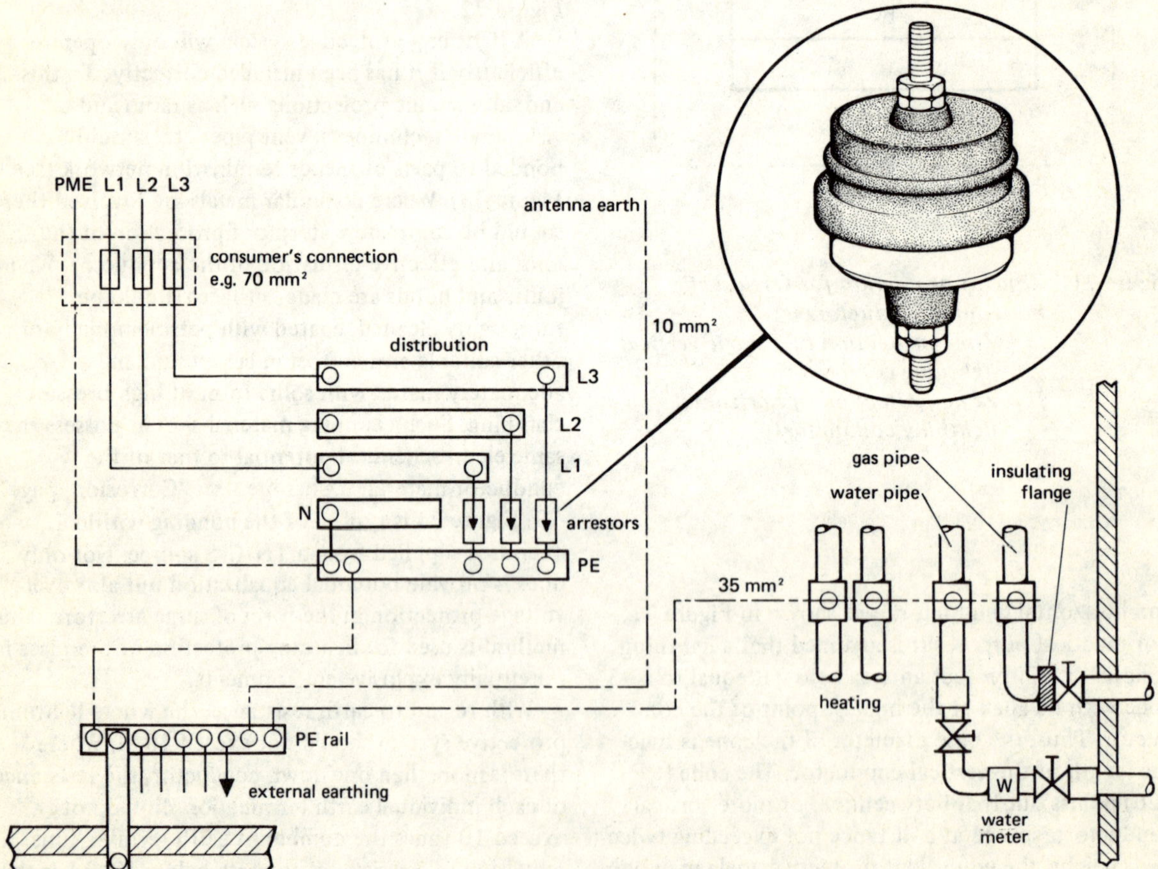

PME L1 L2 L3

consumer's connection
e.g. 70 mm²

antenna earth

10 mm²

distribution

L3

L2

L1

N

arrestors

PE

35 mm²

gas pipe

water pipe

insulating
flange

heating

PE rail

external earthing

water
meter

W

foundation earth

Figure 74   *Internal bonding of a premises*

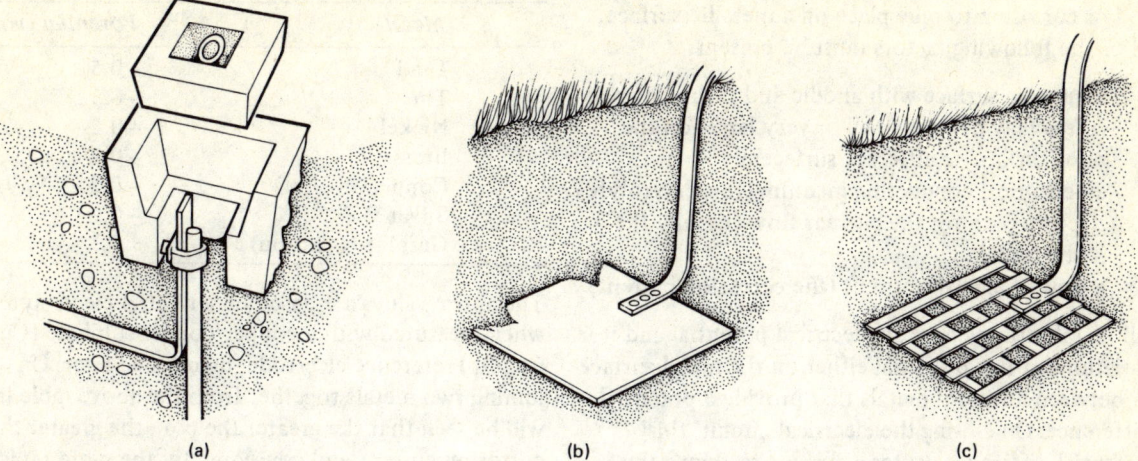

Figure 75   *Methods of connection with the general mass of earth*
(a)   Rod electrode
(b)   Plate electrode
(c)   Lattice electrode

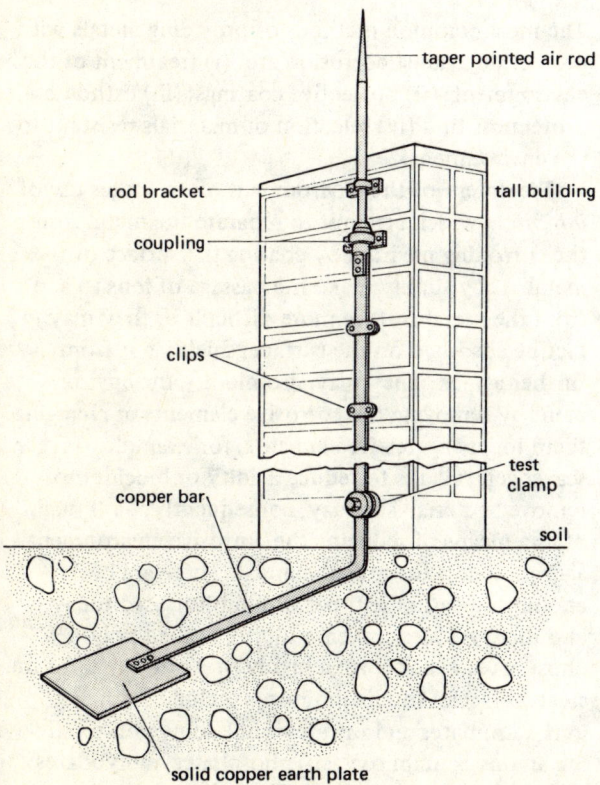

Figure 76   *Components of a lightning protective system*

The tests should be made before any bonding has been effected to metal in or on the structure or to services below ground. Figure 75 shows various types of earth electrode used while Figure 76 illustrates some of the components which make up a typical installation. Earth terminations and electrodes are normally copper or copper alloys except that steel core rods may be used for earth electrodes.

The code recommends that the lightning protective system requires periodic inspection and testing at intervals not exceeding one year. Tests should confirm that the earth resistance does not exceed 10 Ω and that the continuity of down conductors and across joints is satisfactory. Inspection should reveal that the system is mechanically sound and free from corrosion and that no structural alterations have been made to impair the efficiency of the system. CP 326 also recommends that records be kept on site by a responsible person. Such records should show: 'as installed' drawings of the lightning protective system, dates of inspections and tests together with results, details of soil or special earthing arrangements, alterations and repairs as well as information regarding the system's upkeep and responsibility.

Corrosion

This may be described as the chemical or electro-chemical reaction of a metal with its environment resulting in the metal's progressive degradation or destruction whereupon it reverts back to its natural oxide state.

For corrosion to take place on a metallic surface, all of the following factors must be present:

1 a metallic surface with anodic and cathodic areas (these may be very small or very large, on the same surface or different surfaces)
2 an electrical conducting medium (the electrolyte)
3 a return path for the current flow (usually the metal itself)
4 a flow of electric current (the corrosion current)

All metal surfaces have an electrical potential and it is variations of this potential either on the metal surface or between different metals that provide a potential difference. On closing the electrical circuit, this potential difference causes a current to flow – the corrosion current. This flow is from the high potential (anodic) area to the low potential (cathodic) area. These differences in potential are created by differences on the metal surface, flaws, impurities, heat affected zones, physical damage, different metals, variations in concentration of electrolyte, oxygen content, acidity, resistivity changes etc.

The electrochemical reaction is such that, at the anodic area, an oxidation process occurs resulting in a loss of electrons, releasing positively charged metal ions to go into solution. In order for this anodic reaction to take place, a simultaneous reduction process must occur at the cathodic area resulting in a gain of electrons. Without using chemical formulae to explain this reaction effect more clearly, it will be sufficient to say that the flow of corrosion current causes metal ions to leave the surface at the point of discharge and corrosion of the metal takes place. The amount of metal removed is directly proportional to the magnitude of current flow.

The above basically describes a corrosion cell and, as mentioned, all metals have an electrical potential. These can be listed as a galvanic or potential series from the most active or anodic metals to the most passive or cathodic as shown below.

	Metal	Potential (volts)
Active	Magnesium alloy	– 1.6
	Zinc	– 1.1
	Aluminium alloy	– 1.05
	Steel (clean and shiny)	– 0.5 to – 0.8
	Steel (rusted)	– 0.2 to – 0.5
	Iron	– 0.5

	Metal	Potential (volts)
	Lead	– 0.5
	Tin	– 0.3
	Nickel	– 0.2
	Brass	– 0.2
	Copper	– 0.2
	Silver	– 0.1
Passive	Graphite (Carbon)	+ 0.3

The above values are based on a neutral electrolyte when measured with a copper/copper sulphate (Cu/$CuSO_4$) reference electrode, mentioned later. On joining two metals together from the above table it will be seen that the greater the p.d., the greater the corrosion current and consequently the more rapid corrosion to the active metal. This can be seen in practice with the combination of zinc and carbon used in a standard dry cell battery.

Protection against corrosion

The most common methods of providing metals with protection against corrosion are: (i) treatment of the environment; (ii) protective coatings; (iii) cathodic protection, and (iv) selection of materials resistant to the environment.

Treatment of the environment often makes use of *inhibitors* which attempt to separate the metal from the corroding medium by coating the surface of the metal. They simply make the passage of ions to and from the metal surface more difficult or they may in fact be absorbed on the surface, shielding it from further attack. They may also modify the environment by removing the corrosive elements or changing them to less aggressive elements, for example, oxygen scavengers, alkalis to reduce acidity or biocides to remove bacteria. The may, consequently, be thought of as a means of reducing the corrosive environment's aggressiveness towards the metal. There are both anodic and cathodic types of inhibitor used today and there are also mixed inhibitors called polyphosphates which form a highly protective deposit on steel surfaces. Anodic inhibitors include oxidizing ions, chromates and nitrates while some non-oxidizing anions contain oxygen, phosphates, molybdates, silicates and benzoates etc. These function in solutions that are neutral or alkaline and have the effect of causing adjacent anodic areas to become less passive (that is, become less reactive) under the influence of large cathodic currents. Anything that interferes with the cathodic reduction reaction is called a cathodic inhibitor, such as magnesium and calcium salts. These particular ones have low-solubility hydroxides which precipitate in the cathodic regions of the metal and act so as to retard oxygen reduction.

Protective coatings may take the form of painting, organic coatings, chemical coatings or even metal coatings. Painting of metal surfaces not only protects them against corrosion but also makes them visually attractive. Surface preparation is most important and needs to be dry and free from scale. Paints do not provide complete protection on ferrous metals and one should always use a primer first, such as red lead. To some extent, red lead behaves like an inhibitor since the presence in its film of both oxygen and lead soaps (compositions of metal ions and organic acid molecules) act so as to inhibit corrosion. There are many types of paints, primers and coatings all serving the function of providing a resistance to ionic movement. Care must be taken in the selection of primers and coating systems to ensure that they are compatible with each other and with the environment. Further information on corrosion prevention for ferrous materials can be found in British Standard CP 2008 'Protection of iron and steel structures from corrosion'. Chemical coatings might include the phosphating of iron and steel, and the chromating of aluminium, zinc and cadmium. These and other methods provide an appropriate protective surface but afterwards require the application of paint or plastic coating. With metal coatings, steel is often zinc coated in a dipping process called *galvanizing* or it can be heated in zinc dust, a process called *sherardizing*. Steel can also be coated with tin by electroplating it or it can be coated with lead to give it a much higher resistance to atmospheric corrosion. Metal coatings not only improve the corrosion life of the underlying metal but they can also provide low electrical resistance contact as with gold or silver plating; they can provide high reflectivity as with chromium plating and also an oxidation resistance as with aluminium on iron. Finally, it should be remembered that surface preparation is one of the most important aspects of any coating system.

Cathodic protection is a very well known method of corrosion control for buried or immersed structures where the metal to be protected is made wholly cathodic by the artificially introduced anode. This anode provides a protective current which flows in opposition to any 'natural' corrosion current. Complete protection will only be achieved when the magnitude of the anode current balances or exceeds the corrosion current. Figure 77(a) is a simple corrosion cell which in fact could be the anodic and cathodic sites on the surface of a metal structure. To show the principles of cathodic protection, Figure 77(b) superimposes on the simple cell, an external current through an additional anode which sends current through the electrolyte to overcome the corrosion current of the original anode. This stops the original anode from corroding since it now becomes a cathode.

Basically, there are two distinct systems of cathodic protection in use today, they are: (i) *sacrificial anode systems* and (ii) *impressed current*

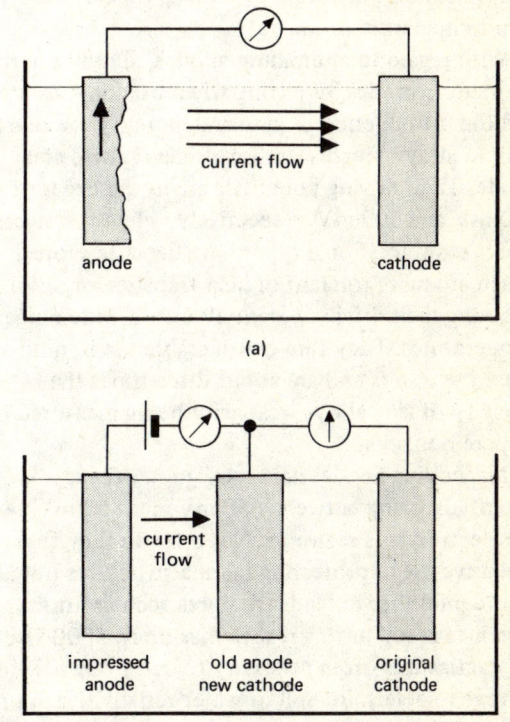

Figure 77 (a) *Corrosion cell*
(b) *Corrosion cell with impressed current to create cathodic protection*

systems. Both are often used in combination with some form of insulation coating on the metal structure they intend to protect: this provides a first line defence against corrosion. In sacrificial anode systems, the anodes in operation are consumed over a period of time. They are always made from less noble metal than the structure they protect. Commonly used anodes (often called 'galvanic anodes') are alloys based on zinc, aluminium, and magnesium. Zinc types are particularly useful in seawater applications for protecting ships' hulls as well as for protecting pipelines where the effects of burial and a high-duty coating tend to give low anode current densities. In general, their driving potential against protected steel is in the order of 250 mV based on normal seawater ambient temperatures. This relatively low driving potential is insufficient to cause damage to most paint coatings and is to some extent self-controlling. Unfortunately, it has a tendency to diminish with increasing temperatures of about 60°C and this limits its application, particularly for metal pipelines carrying high temperature products.

With regard to aluminium anodes, aluminium itself normally corrodes by pitting when used in seawater, and the introduction of zinc and mercury, or zinc and indium alloys, creates a more efficient corrosion anode. Their driving potentials are in the order of 250 mV and 300 mV, respectively, when the electrolyte is seawater. Some types have been developed which are more tolerant of heat transfer conditions, retaining their driving potentials over a wide range of temperatures. They find considerable use in mud and estuarine waters and are about three times the capacity of zinc alloys – capacity being measured in ampere hours/kg.

Magnesium anodes have very good driving potentials, being between 650 mV and 850 mV when the electrolyte is seawater. Not only do they find extensive use in protecting marine structures but also considerable use in land structures such as buried pipelines in soil having resistivities up to 5000 Ω-cm. In practice it is often necessary to carry out site surveys to determine soil or water resistivities in order to ascertain the degree of corrosion. Roughly, severe corrosive soil is up to 1000 Ω-cm whereas moderately corrosive soil ranges between 1000 and 10,000 Ω cm. Above this latter value the soil resistivity is only slightly corrosive. It is perhaps worth noting that

alluvium and light clays have resistivities around 500 Ω-cm and porous limestone ranges between from 3000 Ω-cm to 10,000 Ω-cm. Obviously, the lower the soil resistivity the more economical will be the cathodic protection installation since the current output is governed by the driving voltage and the resistance of the anode circuit.

Figure 78 shows a typical magnesium anode prepacked with *backfill*. This is a low-resistance, moisture-holding material consisting of powdered gypsum (75 per cent), granular bentonite (20 per cent) and sodium sulphate (5 per cent). It is used for the purpose of increasing the effective area of contact

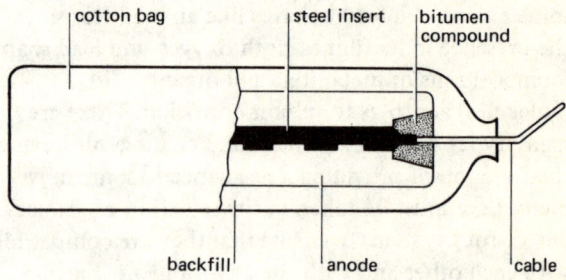

Figure 78   *Magnesium anode*

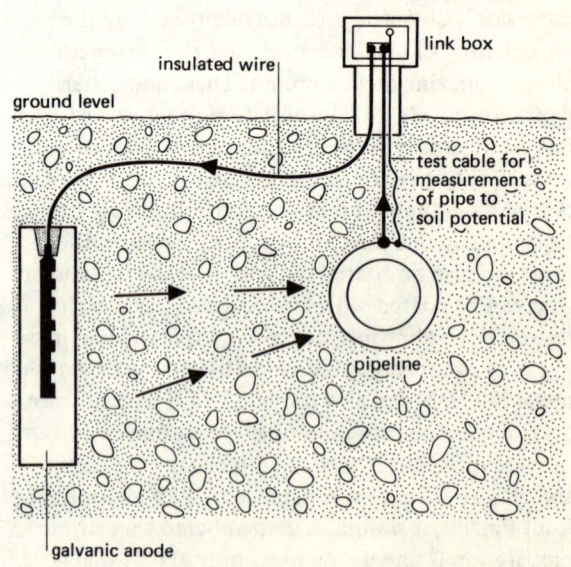

Figure 79   *Sacrificial anode protection for a buried pipeline*

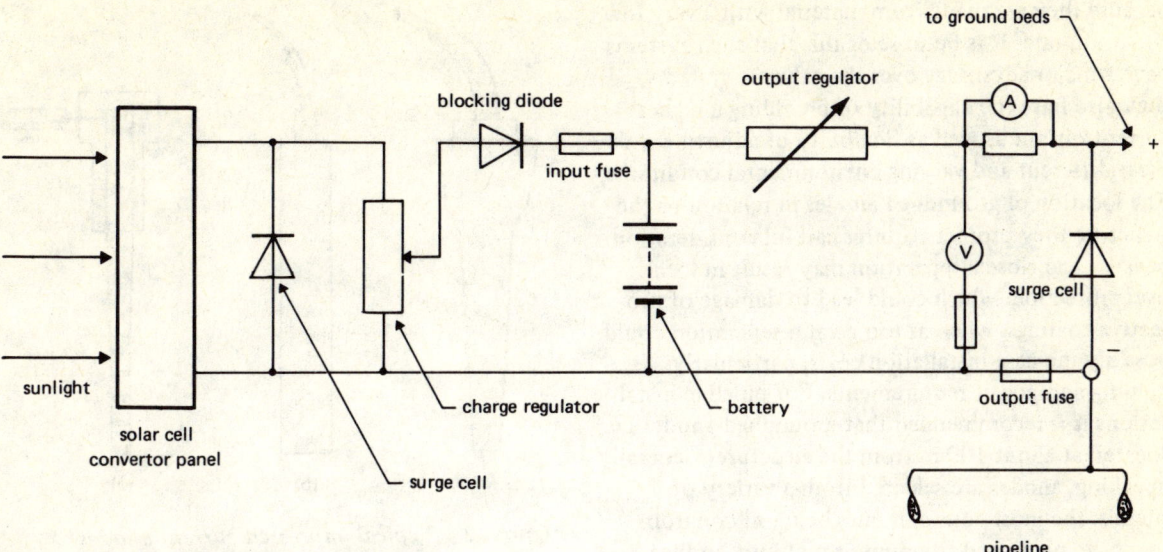

Figure 80   *Solar cell generating system*

with the soil and is contained in a cotton bag or it may be provided loose for site application. Inside the anode will be seen a *steel insert* which serves to maintain the anode's continuity and mechanical strength as it nears the end of its useful life. Attached to the anode is an insulated cable for connection onto the metal structure. An example of this arrangement is shown in Figure 79 where it will be seen that the cable is taken through a *link box* before it is connected to the buried pipeline. The link box allows current measurement to be taken as well as confirming that the anode is operating satisfactorily. The box also contains a test lead attached to the pipeline to enable pipe to soil potentials to be measured and to determine the level of protection. It is normal practice to adequately space the sacrificial anodes along the route of the pipeline, installing them in areas of low resistivity and preferably at a minimum distance of 1.5 m from the pipe. Where the soil resistivity is high, that is, exceeds 500 Ω-cm, then a backfill is spread around the pipe.

Impressed current systems operate by means of a direct current souce obtained from an a.c. system through a transformer rectifier unit. Alternatively, they may be fed from diesel generators, thermo-electric generators, wind generators or even solar panels which find particular use in remote desert locations, protecting oil wells and pipelines from

corrosion. A typical circuit for a solar generator is shown in Figure 80. It has incorporated in it blocking diodes which prevent discharge of the battery through the solar panels in weak light or at night, also surge cells for providing protection against lightning.

The basic feature of Figure 81 is that positive current is impressed via each of the anodes which are laid in ladder formation onto the structure. The negative cable from the pipeline provides the return path. In this way the resulting current flow in the environment protects large areas of the structure with low consumption of the groundbed anodes. This is

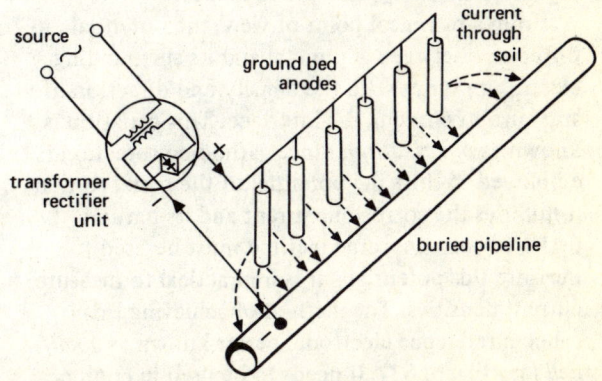

Figure 81   *Typical impressed current system*

because they are made from material with a very low
corrosion rate. It is because of this that such systems
have a major advantage over the galvanic systems;
they also have the capability of providing a higher
current output as well as flexibility of adjustment to
meet different and varying environmental conditions.
The location of groundbed anodes in relation to the
structure they protect requires careful consideration
because too close a separation may result in local
overprotection which could lead to damage of protective coatings, whereas too great a separation could
possibly increase installation costs, particularly
cabling and power requirements. For pipeline installations it is recommended that groundbed anodes be
located at about 100 m from the structure. Generally
speaking, anodes are selected from a variety of
metals, the most common ones being silicon iron,
graphite, platinized titanium or niobium, and lead
alloys. Silicon iron anodes are cast-iron alloyed containing approximately 14.5 per cent silicon. For seawater application the anodes may comprise 4.5 per
cent chromium. Graphite anodes have particular use
in high resistivity environments where anode current
densities are not a determining factor. They are im-
pregnated with linseed oil to reduce their porosity.
Platinized titanium anodes are extensively used in seawater applications with voltages up to 8 V but
platinized niobium can be used for higher voltages up
to 120 V. Both types have very low wear rates with
the niobium being more expensive to produce. Lead
alloys find use in saline and seawater situations,
relying on the formation of a lead peroxide film to
create a conducting medium that does not wear
rapidly: they are unsuitable for mud environments.
Figure 82 shows typical anode shapes.

From a technical point of view, the potential
difference between any metal and its surrounding
electrolyte varies with the density and direction of
any current crossing the interface. This variation is
known as *polarization*. Since cathodic protection is
employed to shift the potential of the metal so that
it nullifies the corrosion current and its natural
desire to corrode, some method must be used to
measure this potential – it is impractical to measure
current densities. The method of achieving this is
called a reference electrode, better known as a *half
cell* (see Figure 83). It needs to be used in conjunction with a high resistance voltmeter, although types

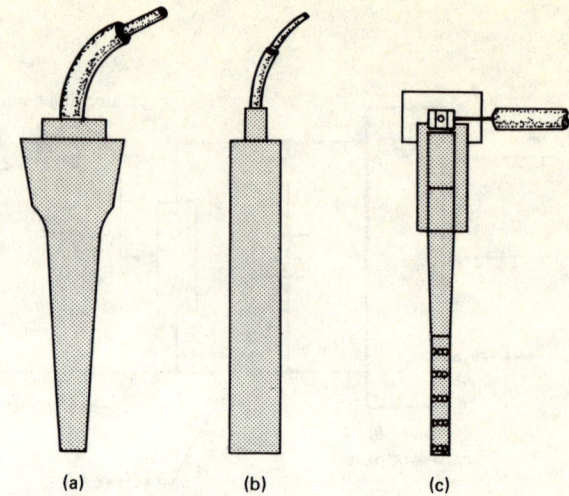

Figure 82   *Typical impressed current anodes*
             (a)   Silicon iron
             (b)   Graphite
             (c)   Platinized titanium

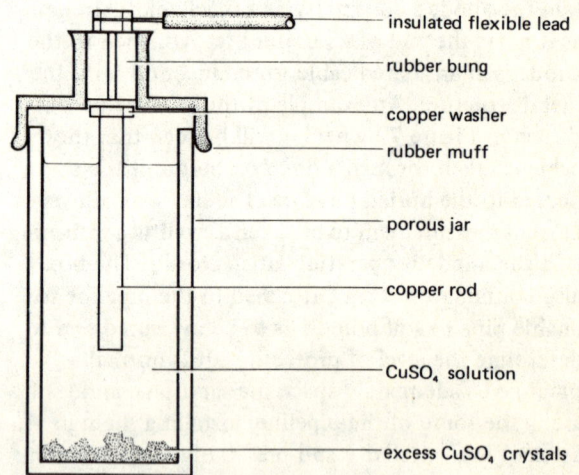

insulated flexible lead
rubber bung
copper washer
rubber muff
porous jar
copper rod
$CuSO_4$ solution
excess $CuSO_4$ crystals

Figure 83   *Copper/copper sulphate reference
             electrode/half cell*

are available which give a direct reading. See Figures
84 and 85. In practice it has been found that a steel
pipeline buried in soil comprising different salts can
have a potential varying from 0.45 V to 0.74 V when
measured against a calomel half cell, that is, a refer-

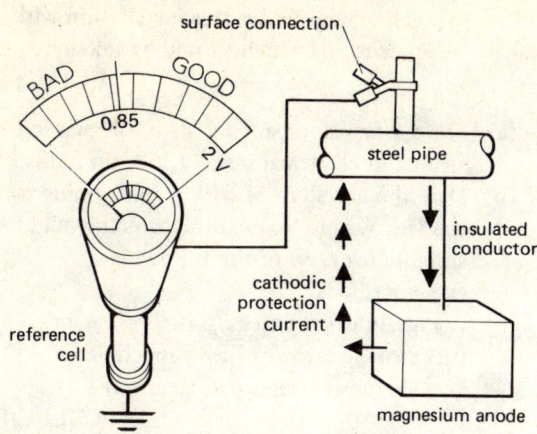

Figure 84   *Direct reading corrosion meter*

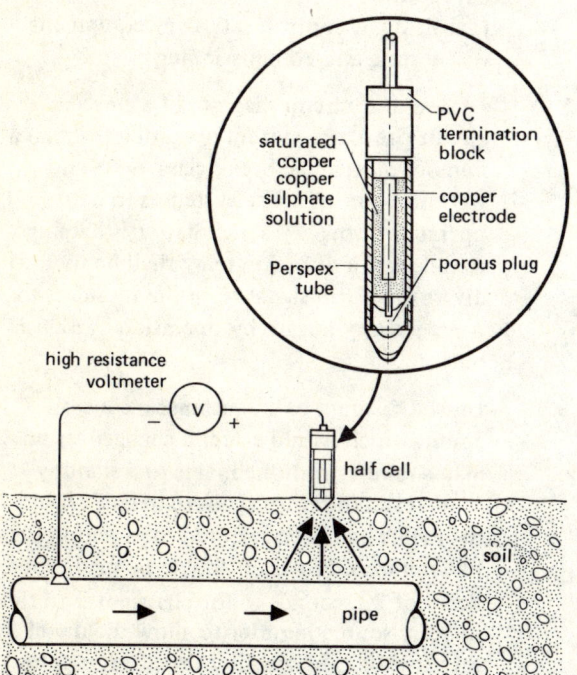

Figure 85   *Method of measuring pipe to soil potential*

(Cu/CuSO₄) for potential measurements of land structures; (ii) silver/silver chloride (Ag/AgCl) for aqueous electrolytes such as seawater, and (iii) zinc (Zn) for clean water and circulating waters. Comparison figures can be obtained for all three types. For example, if the corroding electrolyte is seawater and the metal being protected is cast-iron or carbon steel, then the half-cell values for minimum protected potentials would be -0.85 V relative to Cu/CuSO₄, -0.8 V relative to Ag/AgCl and $+0.25$ V relative to Zn. Any potentials more negative than these normally indicates effective cathodic protection.

The decision as to the current requirements for a particular installation is generally made on the basis of past experience with other structures of a comparable nature. One method is to allow 15 mA/ft² for bare steel in seawater and for coated pipeline in soil it is usual to take 2 per cent of the total pipeline surface area as bare metal and multiply this value by 2 mA/ft². The most important characteristic of an anode material is the relationship between the rate of consumption and the current output. Silicon iron, for example, has current densities between 10 and 50 A/m² in seawater with a consumption rate of 0.5 kg/ampere year. Platinized titanium on the other hand, again in seawater, has current densities up to 500 A/m² with a platinum consumption rate of only 8 mg/ampere year.

Finally, a brief mention of material selection. Careful selection of materials must be made by experienced corrosion engineers or metallurgists, comparing the ideal material for corrosion protection with mechanical, structural and machining properties,

ence electrode consisting of mercury in a standard solution of potassium chloride and saturated with mercurous chloride. This type of half cell is not normally used in field operations owing to its fragility. The usual ones found are: (i) copper/copper sulphate

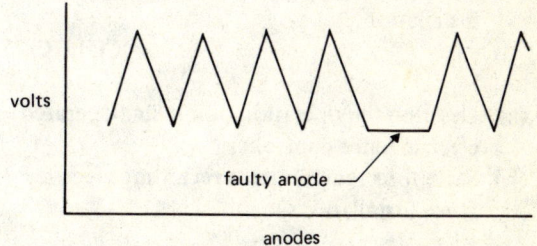

Figure 86   *Graph indicating state of anode groundbed Method is to place a half cell in the soil at a distance from the anodes and to move a second half cell along the surface above the anode system*

costs and many other parameters. Invariably a compromise selection is made for each application from the multitude of metals and alloys available.

The above is a brief introduction to corrosion and corrosion protection systems. It is a specialized branch of engineering and expert advice should be obtained in applying any of the above systems.

Revision exercise 3

1 (a) Distinguish between a Zone 1 and Zone 2 potential explosive area as defined in BS 5345: Part 1: 1976.
   (b) Briefly describe *one* method of protection for electrical apparatus used in a Zone 0 area.
   (c) What is meant by the terms:
        (i)   flashpoint
        (ii) temperature class
        (iii) gas grouping

2 (a) State *four* main requirements which must be met in the wiring installation of a petrol service station.
   (b) Draw the complete line diagram for a two pump installation in a petrol service station where single-motor pumps are used.

                                            CGLI/C/82

3   A pipe-ventilated three-phase motor and starter are to be installed in an explosive atmosphere.
   (a) Describe the installation.
   (b) Sketch the installation to show the ventilation system and ancillary equipment.
   (c) Draw a wiring diagram of the complete installation.

                                            CGLI/C/77

4 (a) Describe the operation of a mains-operated electric fence controller.
   (b) State *four* requirements regarding electric fence installations.

5   A 415 V three-phase and neutral supply is to be taken from a farm building to a milking unit 50 m away. The supply is to be used for driving a 5 kW motor, 10 kW sterilizing unit and 2 kW of dairy lighting. Recommend suitable types of wiring systems, switchgear and protective devices for this installation with all precautions which should be taken.

                                            CGLI/C/79

6 (a) State *four* principal hazards to be guarded against in electrical installations on farms.
   (b) Describe a system of wiring conforming to the IEE Wiring Regulations which would be suitable for *each* of the following
        (i)   cowsheds
        (ii) grain drying areas
        (iii) storage area for heavy machinery (harvester, elevator, tractors)

                                            CGLI/C/81

7   Write brief notes on the following with regard to fire alarm systems:
        (i)   circuit design
        (ii) zones
        (iii) siting of control/indicating equipment
        (iv) testing and commissioning

8   Draw a neat circuit diagram of a closed circuit fire alarm system operating through a number of manual 'break glass' units and warning sounders. The system is to be operated from a 24 V d.c. battery charging cubicle and a diversion relay shall be used to divert the alarm signal from the sounders to a supervisory buzzer by operation of a push button.

9   Draw a diagram of an automatic control circuit which would effect a changeover on mains failure from the mains to a standby generator and return again to mains, after the mains had been restored for a period of two minutes. Provision should be made for a delay of 20 seconds in the changeover to the standby source in order to allow the diesel set providing the standby time to run up to full speed.

                                            CGLI/C/77

10   Figure 87 is a typical office and TBA sales associated with a petrol filling station installation. Using manufacturers' catalogues, make a list of all the electrical items that need installing quoting relevant catalogue numbers.

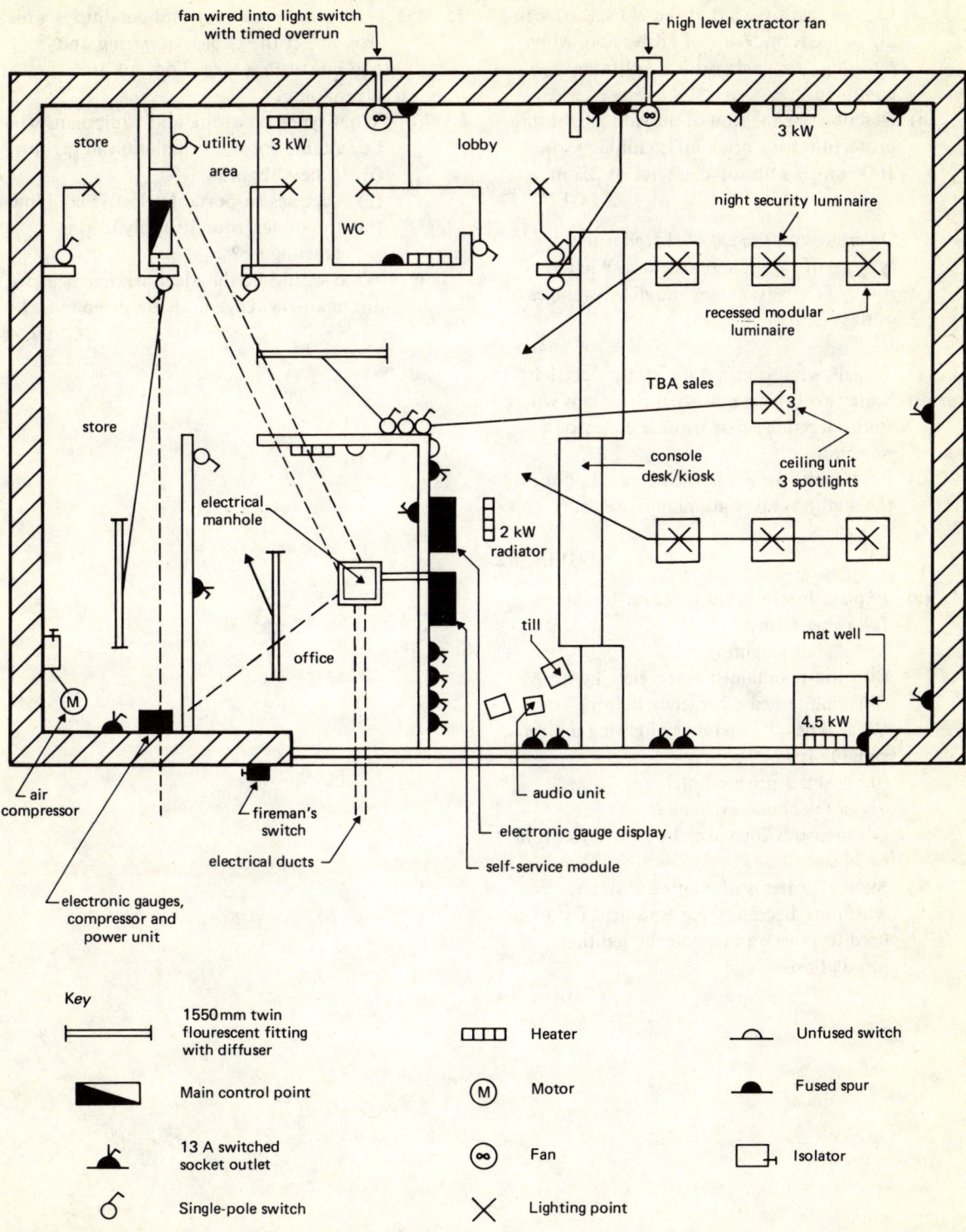

Key

	1550 mm twin flourescent fitting with diffuser		Heater			Unfused switch
	Main control point	(M)	Motor			Fused spur
	13 A switched socket outlet	(∞)	Fan			Isolator
	Single-pole switch	✕	Lighting point			

Figure 87

11 (a) Explain, with the aid of sketches, the mean-
        ing of the term 'Zone of Protection' when
        referring to a single vertical lightning
        conductor.
   (b) Describe one method of providing lightning
        protection for a brick built chimney some
        100 m high with top diameter of 2.5 m.
                                            CGLI/C/77

12   Describe with the aid of sketches the two
        systems of cathodic protection. What are
        some of the advantages and disadvantages of
        both systems?

13 (a) Explain what is meant by 'static electricity'.
   (b) State two typical industrial situations where
        static electricity may cause a dangerous
        situation.
   (c) Explain the danger in *each* case and outline
        the methods taken to minimize static
        electricity.
                                            CGLI/C/82

14 (a) Explain briefly what is meant by the
        following terms:
        (i)   standby lighting
        (ii) non-maintained emergency lighting
        (iii) maintained emergency lighting
   (b) What types of emergency lighting are most
        suitable for
        (i)   a small private hotel
        (ii) a telephone exchange?
        Give reasons for your choice of system in
        each case.
   (c) State *four* items of routine maintenance
        which are necessary for lead-acid batteries
        used to power an emergency lighting
        installation.
                                            CGLI/C/84

15 (a) State *four* environmental conditions which
        may affect the choice of wiring and
        equipment in a large horticultural
        greenhouse.
   (b) What types of wiring and equipment would
        be suitable for the installation in (a) for;
        (i)   general lighting
        (ii) supplies to portable electric equipment
        (iii) the underground supply to the
              greenhouse?
   (c) What would be the ideal arrangement for
        the main switchgear at the greenhouse?
                                            CGLI/C/84

chapter four

Lighting and heating systems

After reading this chapter you will be able to:

1 State a number of comparisons between different types of lamp, particularly MCF, MBF, SOX and SON discharge lamps.

2 Describe a number of factors to be considered when dealing with lighting in hospitals and offices.

3 Describe a number of requirements associated with public lighting installations, namely, columns, lanterns, lamps and control gear as well as the disposal of certain discharge lamps.

4 State the difference between direct acting heaters and thermal storage heaters.

5 Explain the operation and control of thermal storage heaters.

6 Distinguish between types of water heating system such as dual element, cistern, non-pressure and instantaneous heaters.

7 Perform calculations to find power required to heat rooms.

Lighting systems

In Volume 2 of this series, Chapter 6 is devoted to installation of lighting, covering types of lamp and their control gear, regulation requirements, emergency lighting, stroboscopic effects and effects of harmonics, and lighting calculations. This chapter will briefly cover some of these points again as well as covering public lighting, hospital lighting and office lighting.

Types of lamp in common use today are shown in Table 4. Figure 88 shows examples of various types of discharge lamp listed by their letter codings. Tungsten halogen lamps are seen grouped with tungsten filament lamps mainly because they have descended from them. In practice, filament lamps (GLS lamps) have advantages of low cost, simple operation and good colour rendering (that is, how a lamp renders coloured objects compared with that of daylight). Their efficacy range is relatively low, not exceeding 18 lumens/watt for the highest wattage lamp available. This, and a shorter life expectancy

than most other lamps, limit their use to domestic lighting and other lighting where no real importance is placed on selection. Tungsten halogen lamps, on the other hand, have several advantages over the GLS

Table 4   *Types of lamp in common use*

Filament Lamps (GLS) and tungsten halogen lamps	filaments
Low pressure tubular fluorescent lamps (MCF)	phosphors
Low pressure sodium lamps (SOX and SLI)	
High pressure sodium lamps (SON, SON–T, SON–TD, and SON–R)	
High pressure mercury lamps (MB, MBF, MBFR)	discharge
High pressure mercury blended lamps (MBTF)	
High pressure metal halide lamps (MBI and MBIF)	

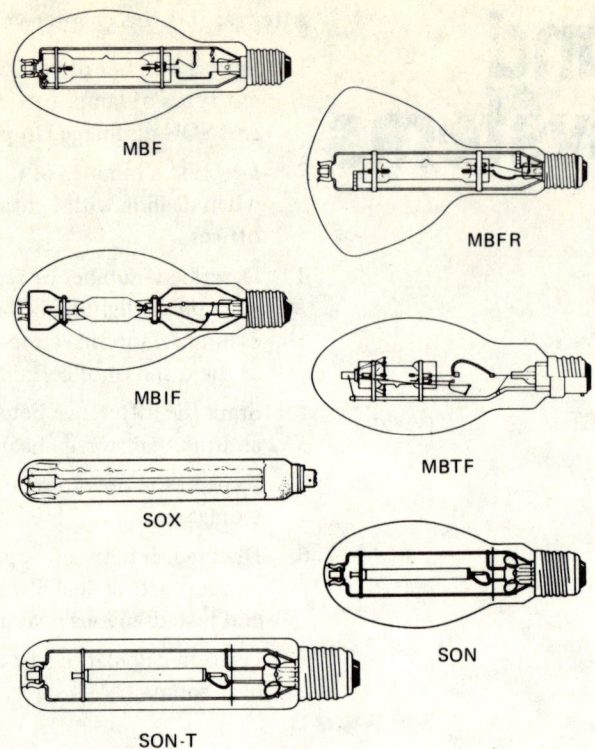

Figure 88   *Common discharge lamps*

lamps with an approximate doubling of efficacy values (reaching around 30 lumens/watt), longer life expectancy and 100 per cent lumen maintenance. Their application is mainly for floodlighting, display lighting, traffic lights, projectors and vehicle head-lamps. Future development might easily see them being used in domestic dwellings.

Tubular fluorescent lamps, the next consideration, incorporate a low pressure discharge but because the majority of light output is from phosphors, they fall into a category of their own. The most recent develop-ment has been in the use of krypton instead of argon as a gas filling and also in the use of a new range of high temperature phosphors (called triphosphors) which emit light in specific narrow spectral bands throughout the visible range. Krypton-filled fluores-cent lamps are made with 26 mm diameter and 38 mm diameter tubes and should be the first choice for switch-start circuits since they consume 10 per cent less energy than standard tubes, with little or no

loss of lumen output. The multi-coated or poly-phosphor tubes on the market today are designed to give a much improved colour rendering. Efficacy values for triphosphor tubes are about 70 lumens/watt (including control-gear losses) and their nominal life is around 7500 hours. Both these factors vary considerably according to the type of lamp used, the environment in which it is used and the frequency of switching. Fluorescent lamps have wide application and little further will be said about them at this stage.

Discharge lamps, namely, low pressure sodium (SOX and SLI), high pressure sodium (SON/T&TD/R), high pressure mercury (MB/F/FR and MBTF) and metal halide (MBI/F) lamps complete the final group-ing. The light from both the low pressure sodium lamps is concentrated in the yellow part of the spectrum, giving a single-hue colour rendering. Despite this, the efficacy values are extremely high, being the highest of all the lamps mentioned, ranging between 123 and 175 lumens/watt. They find considerable use in road and street lighting and also security lighting. Their life expectancy is in excess of 6000 hours. The high pressure sodium lamps, again, have very good efficacy values, ranging from about 80 lumens/watt to 130 lumens/watt. The SON lamps have a warm, golden colour appearance with a moderate colour rendering index which has made them one of the most successful lamps in the last ten years, replacing a number of other lamps, mainly mercury vapour lamps. The smaller wattage lamps have very long life expectancies, being around 20,000 hours. This makes them a worthwhile proposition for roadlighting, floodlighting and numerous commercial and industrial lighting installations. High pressure mercury vapour lamps of the MBF and MBFR type have a nominal life of 7500 hours. Their efficacy range is around 50 lumens/watt with a cool blue colour appearance and fair colour rendering. MBF lamps have both industrial and commercial use besides shopping centre lighting and concourse lighting. The MBFR lamps also have numerous uses particularly in highbay lighting where their internal reflectors are unaffected by dirt and atmospheric corrosion such that they require minimum maintenance. The high pressure mercury blended lamps (MBTF) have a rated life around 8000 hours and require no control gear. They have relatively low efficacy values (16 to 23 lumens/watt) and find considerable use in places where access is often difficult, for example shop windows, warehouses and street lighting. The metal halide lamps also have a nominal life around 7500 hours which again could be exceeded where infrequent switching takes place. With efficacies between 63 and 72 lumens/watt (including control-gear losses), they are used in situations where a high light output must be combined with good colour rendering, for example, colour television cameras and studio lighting, floodlighting of buildings and football stadia.

The above briefly outlines lamp types and some of their characteristics and uses. It is important in all lighting installations to consider some form of preventive maintenance which serves the purpose of keeping the lighting in good condition. This will involve cleaning luminaires, preventative methods to reduce corrosion, inspection and testing, and periodic lamp replacement. Obviously, the cleaning of the luminaires will depend a great deal upon the degree of atmospheric pollution in their working environ-

ment. Such procedure is recommended to take place every six months in industrial locations but this will of course depend on the working conditions of the industry concerned. There is a view that labour costs can be greatly reduced if the cleaning cycle is phased in with the replacement cycle of the lamps when this becomes due. Also at cleaning times, checks on the luminaires can be made for rust and corrosion. This latter problem can be controlled by spraying any exposed metalwork with a silicone rust-inhibitor after the luminaires have been cleaned. In some industrial premises, condensation is a real problem and can have undesirable effects on the lighting installation. Checks should therefore be made on any hinges or sealing gaskets associated with the lighting equipment and appropriate action taken where faults are found. Where tests of insulation resistance are made, care should be taken not to damage electronic control gear, such as ignitors, with the high testing voltage applied. Such components should either be disconnected from the circuit or shorted out. With regard to lamp replacement, the indications for GLS and tungsten halogen lamps is when approximately 5 per cent of their installed number has failed. Mercury vapour lamps should be replaced after approximately 7500 hours with MBIF lamps around 6000 hours. Sodium vapour lamps are more difficult to assess but it should be realized that such lamps draw more energy from the supply as they age and also give lowering efficacy values.

It is important that replacement lamps are compatible with existing control gear and luminaires. If this is not so, problems will arise with lamp starting, overheating control gear, unstable light output, excessive current drain, noisy operation and likelihood of excessive flicker with a risk of increased *stroboscopic effect* (that is, an effect in which rotating objects can appear to be stationary). Discontinuous light is a source of danger and all discharge lamps have a 'flicker' content in their light output at twice the supply frequency. Lamps, however, with a phosphor coating, such as MCF, MBF and MBFR lamps, tend to produce a steady output which make them less likely to produce stroboscopy. Where this effect is a problem it may be overcome by designing circuits over two or three phases of the supply, using *lead–lag* luminaires, selecting lamps with a low flicker characteristic or increasing the general lighting with

GLS or PAR lamps (that is, pressed-glass reflector lamps). With few exceptions, it is always advisable to re-lamp luminaires with the same make unless a check has been made with both lamp and luminaire manufacturer. This also applies to control gear. It should be pointed out that the mismatching of lamps to control gear can quite often occur in error either at the ordering stage or installation stage. It is important that steps are taken to avoid this happening. It is also important to read manufacturers instructions and warning notices concerning handling and disposal of lamps and equipment.

Hospital lighting

Lighting requirements in a hospital are so diverse that nearly all the products listed in a lighting manufacturer's catalogue could be used. There are, however, certain areas which require extra care when selecting luminaires, particularly the wards and other patients' rooms. In the hospital ward, for example, there must be sufficient light for the medical staff to care for their patients, while at the same time patients require a lighting level that will allow them to rest and sleep without causing any discomfort. Luminaires should be so designed to produce the minimum amount of dust collection on their surfaces, the top, for example, can be made flat and be free from unnecessary protrusions. They should be designed so that they are simple to dismantle for cleaning and maintenance purposes. General ward lighting should be designed so that only a little light reaches the patient's head position and special lighting should be designed to enable medical staff to carry out detailed examination on patients without moving them to another location. For this purpose it may be advantageous to install either additional down-lighter type luminaires over each bed or wall-mounted adjustable luminaires – both being switched at the bedhead position. A down-lighter luminaire would also be desirable over the night nurse's desk, allowing the nurse to read and write reports without disturbing sleeping patients.

The general lighting found in operating theatres and their ancillary rooms is usually provided by surface-mounted or recessed fluorescent luminaires. Operating table lighting is usually by means of a large counterpoised pendent housing a number of low

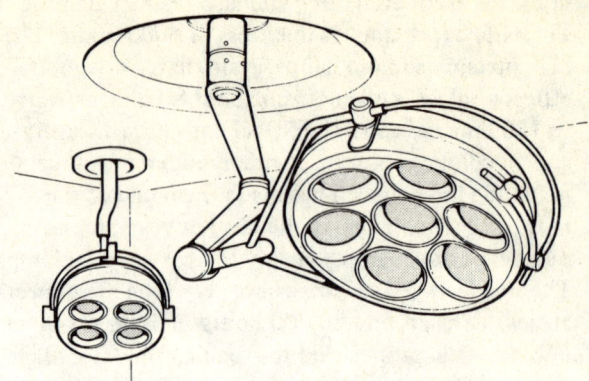

Figure 89   *Typical operating theatre lighting*

voltage lamps in parabolic reflectors. Such luminaires use heat-resisting glass and dichoic reflecting surfaces to reduce radiant heat from the lamps. Since anaesthetics now used are much less inflammable, the risk of explosion is not quite so serious and because of this general lighting levels can be controlled through dimmers. Figure 89 shows a typical operating theatre luminaire.

Office lighting

The majority of conventional offices today with average ceiling heights (say around 3 m) will be adequately lit by fluorescent tubes enclosed in prismatic luminaires. Small offices accommodating only one or two people in an area about 10 m^2 will probably find two single-tube luminaires sufficient where it is recommended that 36 W Polylux tubes be used. It is recommended that a uniform illuminance of 500 lux be achieved and in large offices, where possible, energy savings may be made by using asymmetrical luminaires distributed around the perimeter of the room, which are switched off when daylight conditions are sufficient. Switching control for lighting in large offices is generally in rows or blocks with thought being given to phase balancing of the lighting load.

In large open-plan offices where air conditioning is employed, one sometimes finds air-handling luminaires. These serve the purpose of extracting stale air in the room and keeping the fluorescent tubes cool during operation. In this combined system it is possible to extract the heat from the luminaires and re-use it for other purposes. Office lighting, in general, should present a pleasant and cheerful appearance for both staff and visitor to the premises.

Public lighting

Although public lighting is primarily the responsibility of a municipal lighting engineer, it is not uncommon these days for the private electrical contractor to become involved with the provision of lighting for public areas.

Public lighting covers roads, footpaths, parks, car parking as well as shopping areas and other amenity lighting installations. The 1981 Land and Planning Act has given greater opportunity for the electrical contractor where he can offer not only a much quicker completion of a programme coupled with adequate resources and a range of plant and equipment but also make available a highly motivated workforce which ultimately can only lead to a more efficient and economic service.

A good design of public lighting installation will need to consider a number of factors without necessarily forgetting the aesthetic aspect of making the installation impressive in daylight hours as well as being pleasing in the hours of darkness. It is for this reason that care should be taken in the correct selection of column, lantern and light source used.

In practice, street lighting columns are mainly made of concrete, steel or aluminium. They come in various sizes ranging from 5 m to 15 m. These are shown in the Table 5 in relation to their planting depth. Figure 90 shows typical column structures.

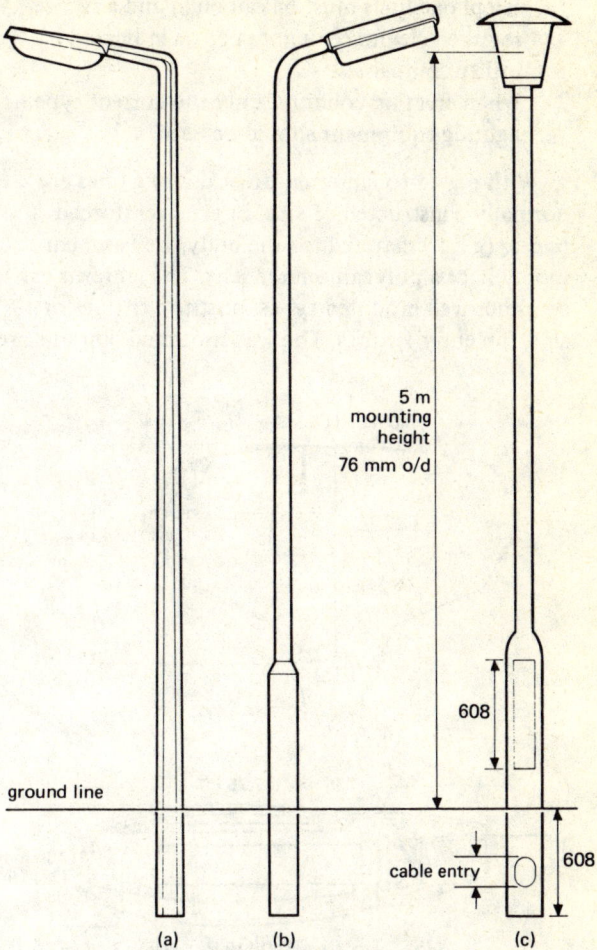

Figure 90   *Sreet lighting columns*
    (a) 5 m concrete column with bracket and side entry lantern
    (b) 5 m steel column with bracket and side entry lantern
    (c) 5 m aluminium column with post-top lantern

Table 5   *Standard street lighting columns*

Nominal height (m)	Planting depth (mm)
5	800
6	1000
8	1200
10	1500
12	1700
15	2000

When erecting a lighting column the following points should be considered:

1   Care must be taken when excavating to avoid the possibility of live cables and pipes being underneath the surface of the ground. It is always wise to check beforehand with the utility companies for information relating to the location of their services.

2   Prior to erection of columns and the fitting of brackets etc., details of overhead electricity and

telephone lines must be obtained and any
required clearance maintained as indicated by the
utility companies.
3   When erecting columns, only the correct type of
lighting equipment should be used.

With regard to lanterns, street lighting ones are
normally constructed of steel or glass-reinforced
plastic (g.r.p.) material for the body shell and commonly have a polycarbonated lens. The lanterns can
be either wall mounted types, post-top fittings or
side/top entry fittings. The wall mounted lanterns are

usually used in locations where it is not possible to
erect a column in say an alley-way. Post-top lanterns
would normally be used for columns erected along a
footpath or in open shopping areas. Top or side entry
mounted lanterns are normally used for street lighting
columns along public highways and a choice of cut-
off or semi cut-off lanterns can be used, depending
upon the area requiring to be illuminated, if it is built
up or open. Figure 91 shows several common types of
lantern used.

The control gear for the lanterns often consists of
the following:

(a)   cut-out fuse unit fitted with an HBC fuse
(b)   time control unit such as a time switch or photo-
      electric cell
(c)   ballast transformer
(d)   power factor correction capacitor
(e)   ignitor (when fitted), see Figure 96

Figure 93 is typical of the control gear found at the
base of a street-lighting column supporting a 400 W

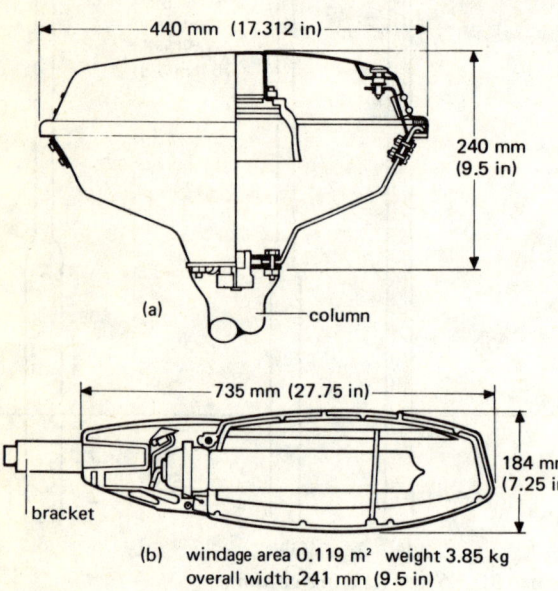

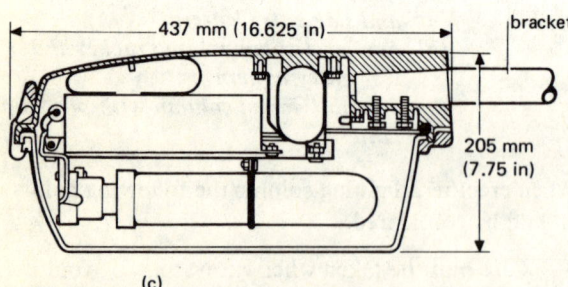

(b)   windage area 0.119 m² weight 3.85 kg
      overall width 241 mm (9.5 in)

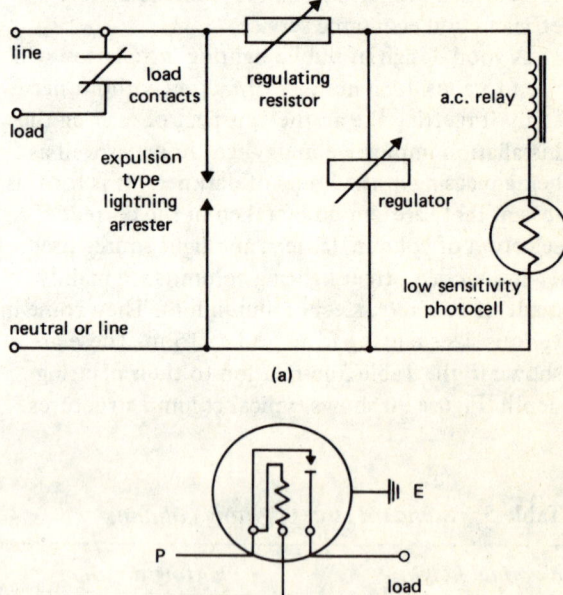

Figure 91   *Types of lantern*
            (a)   Post-top lantern for 80/125 W
            *   MBF/U lamp*
            (b)   Semi-cut off sodium for 90 W SOX
            *   lamp*
            (c)   Side-entry lantern for 35 W SOX
            *   lamp with control gear incorporated*

Figure 92   *Typical time control methods*
            (a)   Circuit for cadmium sulphide
            *   photocell*
            (b)   Wiring for a two-phase switching
            *   cycle time switch*

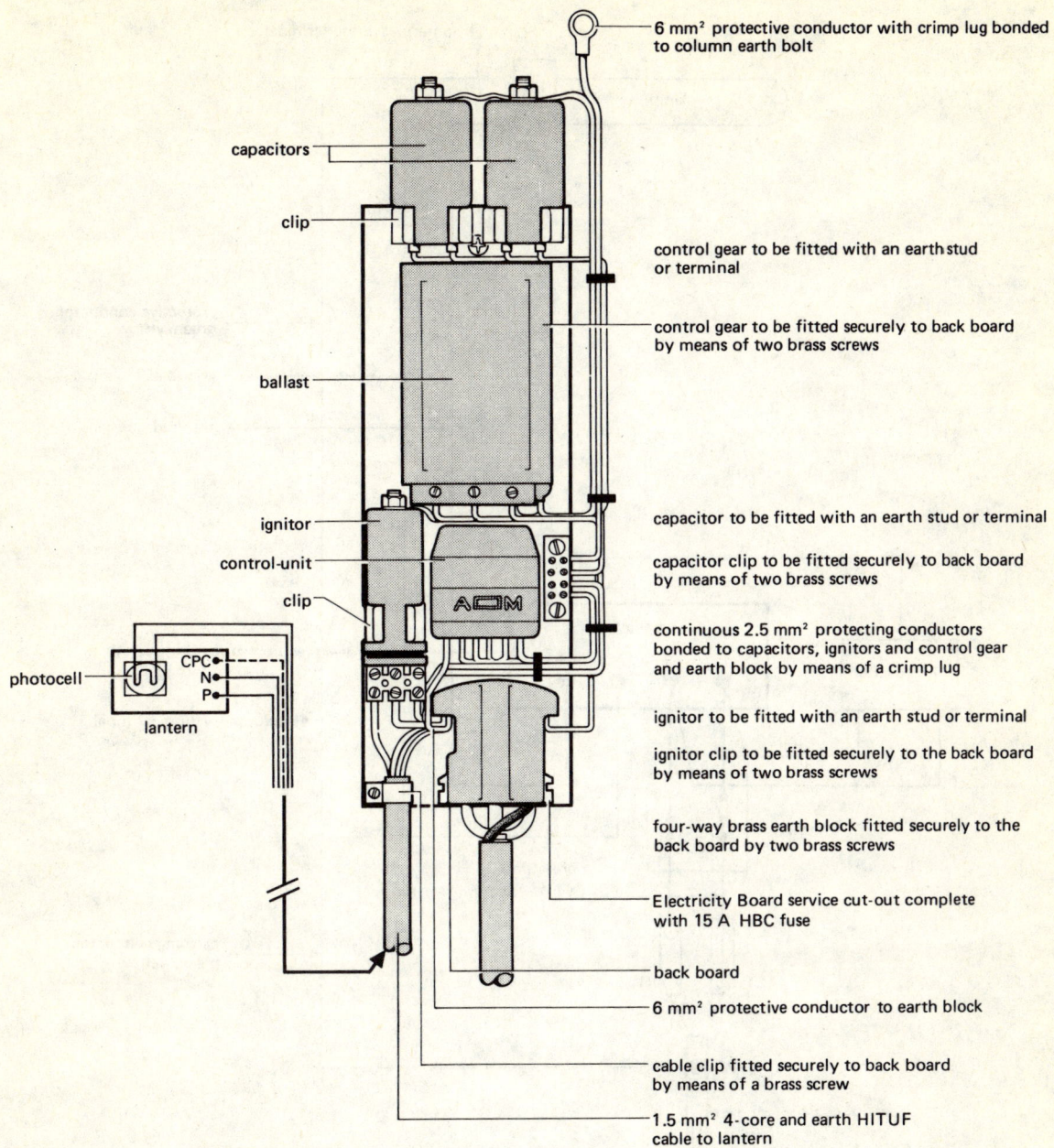

Figure 93 *Method of wiring a lighting column for 400 W SON lamp*

SON lamp. Time control units may be wired to suit individual lanterns or a group of lanterns. Figure 92 shows two arrangements for their control. In diagram (a) it should be pointed out that the load contacts are shown in the normally closed position with no light on the photocell.

In terms of the lamps used, the most common are the SOX, SON and MBF/U. The SOX lamp has been

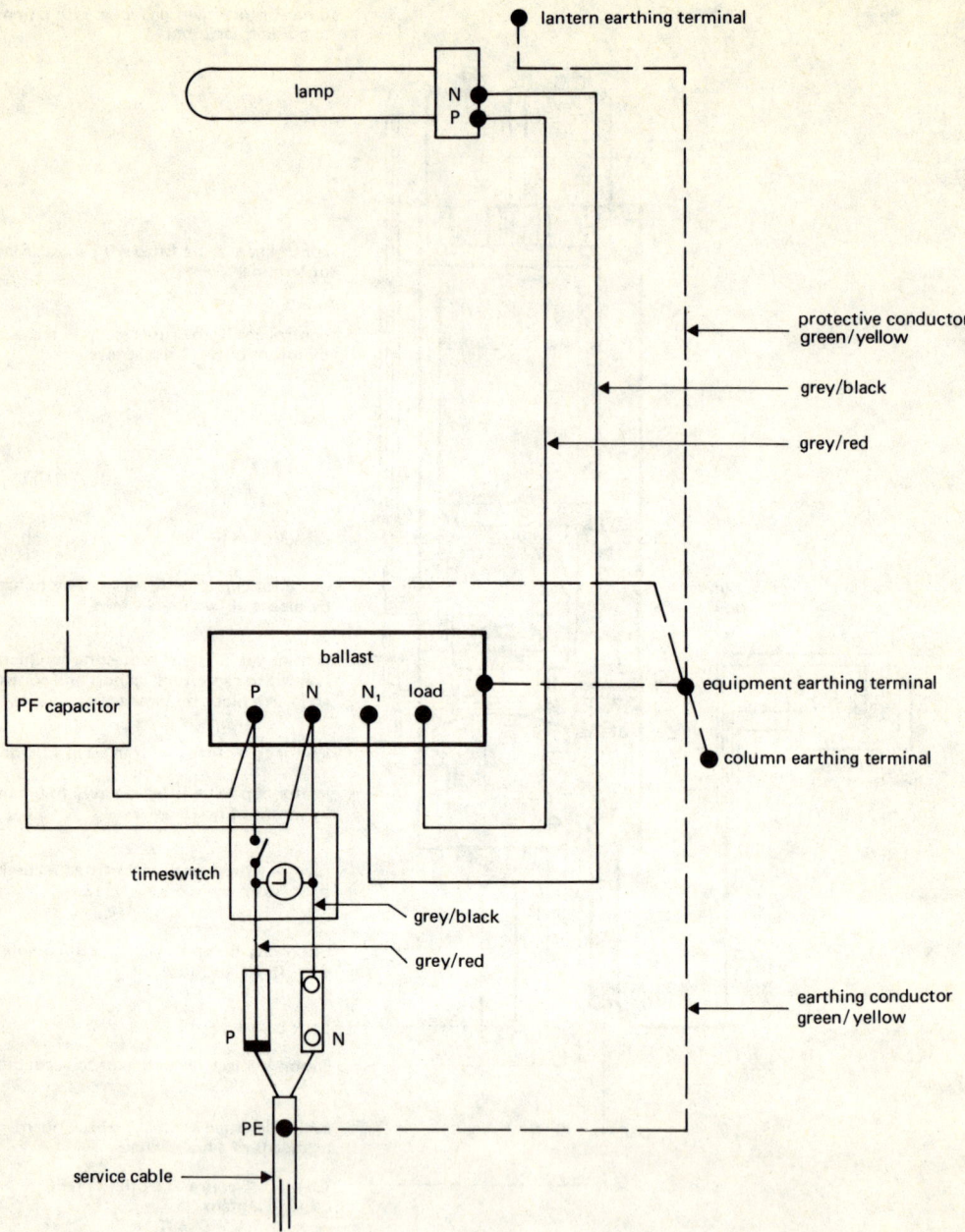

Figure 94   *Typical wiring arrangement for single arm lighting unit with time switch control*

used since 1932 and while being very economical it has suffered from having a poor colour rendering. The SON lamp by comparison is a fairly recent addition and has good colour rendering and can also be used to its eventual destruction. The MBF/U lamp however,

was introduced as an alternative to the SOX lamp because of its better colour rendering but unfortunately its energy consumption is much higher. In practice, lamps used in columns up to 6 m in height are known as Group B and those over this figure

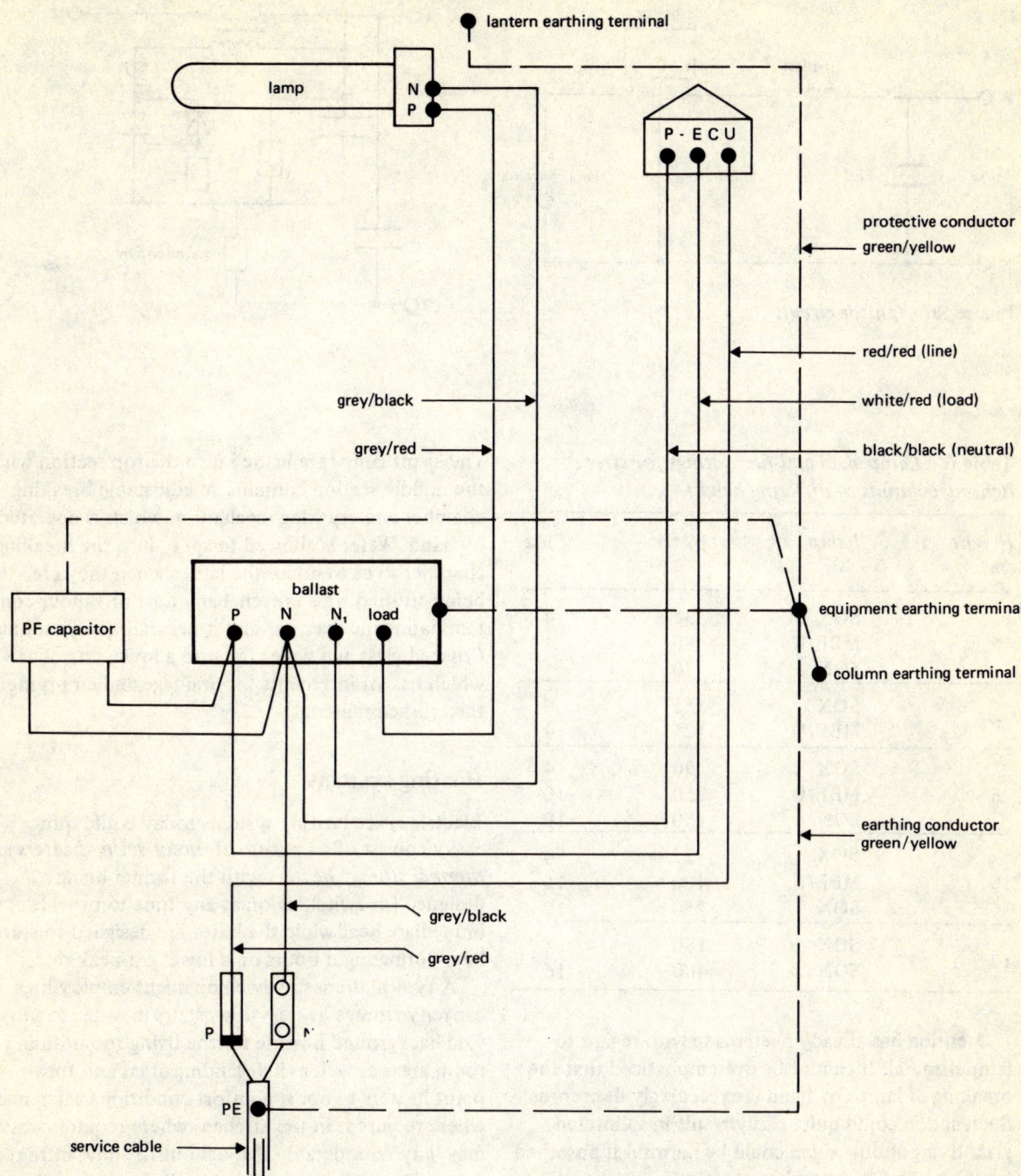

Figure 95   *Typical wiring arrangement for single arm lighting unit with 1-part photoelectric control*

known as Group A. Figures 94 and 95 show the methods of wiring single arm lighting units with their time switch control, and Figure 96 shows an ignitor.

Table 6 shows the relationship between different column heights and types of lamp with their respective wattages and fuse ratings.

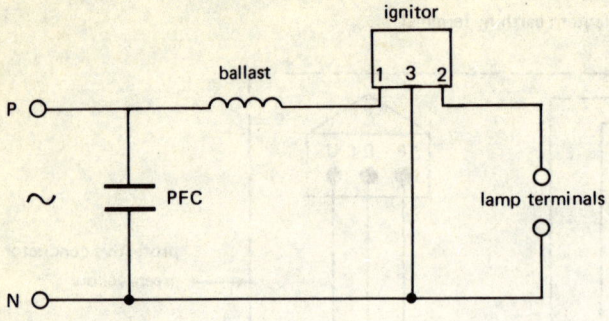

Figure 96 *Ignitor circuit*

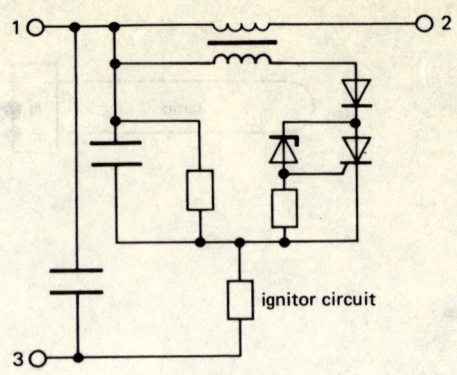

Table 6 *Lamp sizes and fuse ratings for street lighting columns of differing height*

Height (m)	Lamp	Watts	Fuse
5	SOX	35	4
	MBF/U	80	4
	SON	70	4
6	SOX	55	4
	MBF/U	125	4
8	SOX	90	4
	MBF/U	250	10
	SON	150	10
10	SOX	135	4
	MBF/U	400	16
	SON	250	10
12	SOX	180	4
	SON	400	16

Mention has already been made with regard to lamp disposal. It cannot be overemphasized that the breaking of lamps by hand is exceedingly dangerous. Such action could quite easily result in splintered glass flying about which could be harmful if absorbed by the body. The phosphate coating used in mercury lamps could be poisonous if the powder entered an open wound and sodium will ignite with water which could possibly start a fire. There are several lamp crusher units on the market. These often consist of a long stainless steel enclosure having several sections.

The spent lamps are loaded into the top section while the middle section contains an adjustable breaking chamber and crushing mechanism which is operated by hand. Water is allowed to spray into the breaking chamber so as to douse the lamps when they are being crushed: this prevents any dust or vapour contaminating the operator and it neutralizes any sodium. Crushed glass and water fall into a lower chamber which has arrangements for drainage and emptying the crushed material.

Heating systems

Electric space heating systems today could quite easily consist of a mixture of *direct acting heaters* and *thermal storage heaters* with the former being designed for switching on at any time to provide immediate heat while the latter are designed to store heat during night hours on a lower off-peak rate.

A typical domestic dwelling might employ high capacity storage heaters downstairs in order to provide background heating for the living room/dining room areas as well as hall/landing areas and focal point heaters to boost comfort conditions when and where required. In the kitchen, where requirements may vary considerably throughout the day, infra-red heaters or convector heaters are likely choices. Bedrooms could utilize thermostatically controlled oil-filled radiators or again use convector heaters which could be free-standing types or wall-mounted types having their own built-in thermostats. Arrangements can be made to control fixed heaters with a room

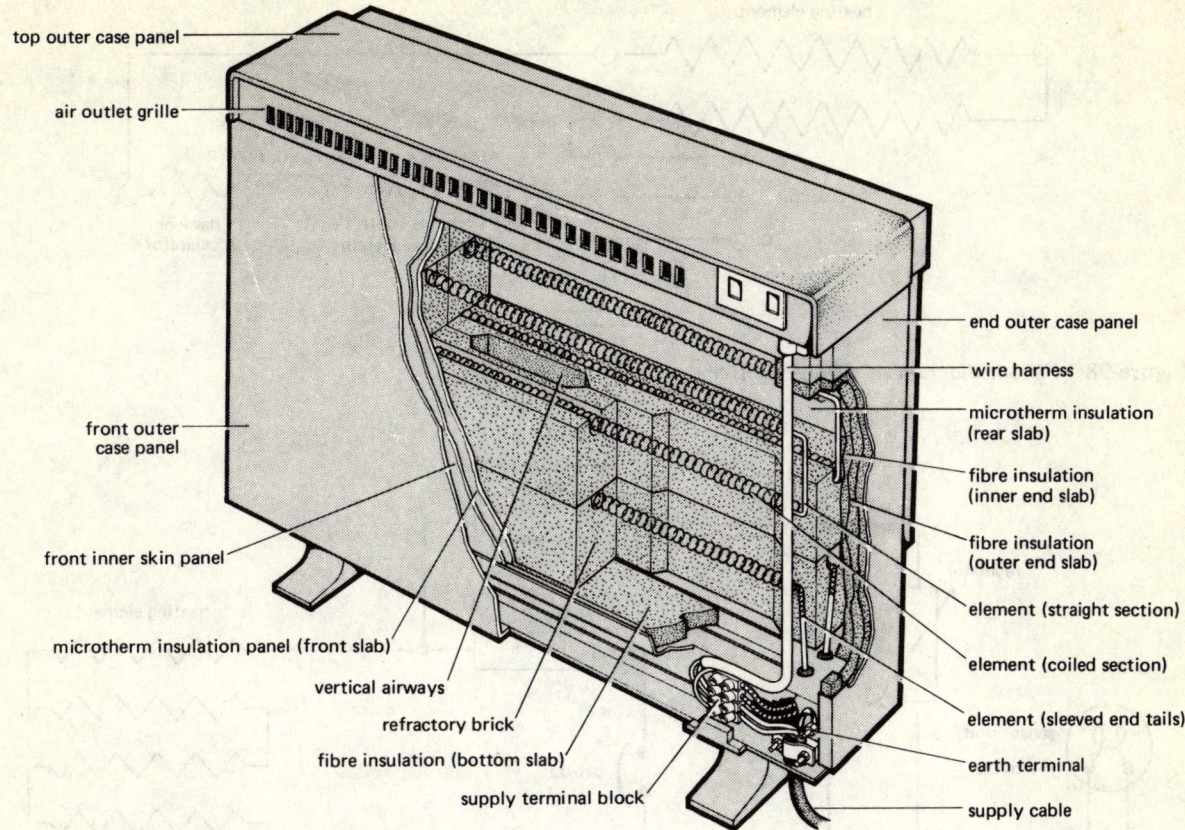

top outer case panel

air outlet grille

front outer case panel

front inner skin panel

microtherm insulation panel (front slab)

vertical airways

refractory brick

fibre insulation (bottom slab)

supply terminal block

end outer case panel

wire harness

microtherm insulation (rear slab)

fibre insulation (inner end slab)

fibre insulation (outer end slab)

element (straight section)

element (coiled section)

element (sleeved end tails)

earth terminal

supply cable

Figure 97 *Internal design of a thermal storage heater*

thermostat via a time clock and White Meter. The heating of a bathroom is often with an infra-red heater or oil-filled radiator and there are combined infra-red heaters and lighting fittings available for ceiling mounting. These are controlled by a pull cord switch providing both heat and light either simultaneously or independently.

The choice of focal point heater and convector heater is considerable and a number of modern designs now have automatic energy controllers. Fan heaters are very popular, particularly the portable models with their two-speed fans which are able to raise the room temperature within minutes; also available are heated towel rails and skirting convector heaters which fit on or in place of the existing skirting.

Thermal storage heaters are in widespread use today, taking advantage of the *Economy 7* off-peak tariff offered by Electricity Boards (that is, a night

rate of 1.9p/unit for seven hours, normally between 1.0 a.m. and 7.0 a.m.).* The heaters depend on the material to store the heat and the thermal insulation to retain it. The firebricks or refractory bricks are often formed from clay, olivine, chrome and magnesite which give a high thermal capacity and conductivity. The thermal insulation used is often in the form of fibres giving blankets of rock wool, mineral wool and glass fibre – *microtherm* is a typical material. Figure 97 is a diagram of a thermal storage heater showing its internal design. Figure 98 is a typical wiring circuit. The charge controller regulates the amount of heat which can be put into the core during the charge period and is mounted on the front, top or side of the heater. At the side of this controller is an output controller which consists of a bimetal actuated damper mechanism. This serves the purpose

*Based on 1983 figures.

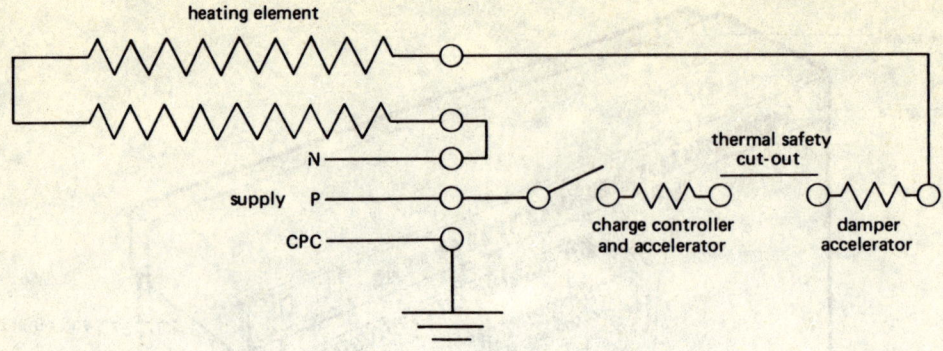

Figure 98   *Circuit diagram of storage heater*

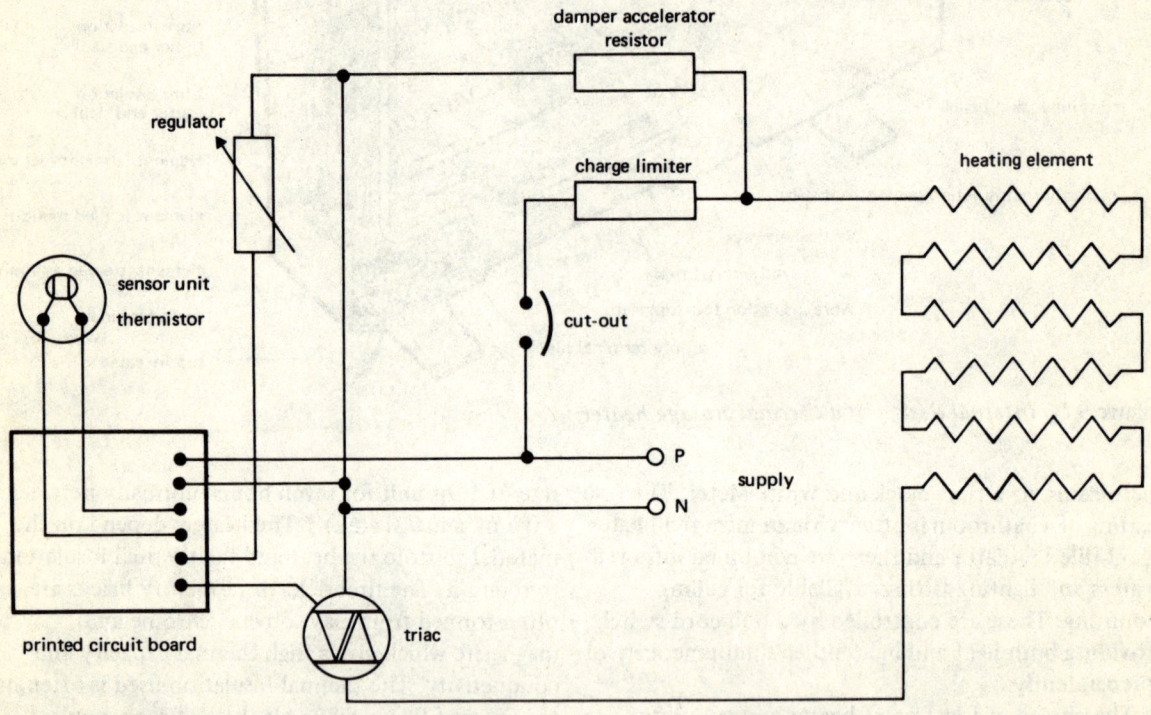

Figure 99   *Electronically controlled storage heater*

of releasing extra heat from the core as and when required. At certain knob control positions this mechanism will operate automatically. A thermal safety cut-out is also shown. Figure 99 is a circuit diagram of an electronically-controlled thermal storage heater and this type uses a remote electronic temperature sensor which is designed to monitor room temperatures during the charge period; it auto-

matically regulates the amount of charge taken in by the heater according to the heat losses in the room. By doing this, the comfort level in the room is automatically controlled at the end of the charge period irrespective of weather conditions. Figure 100 shows a circuit diagram of a fan-assisted type storage heater. The fan is normally controlled by a room thermostat and it draws in room temperature air at the rear of

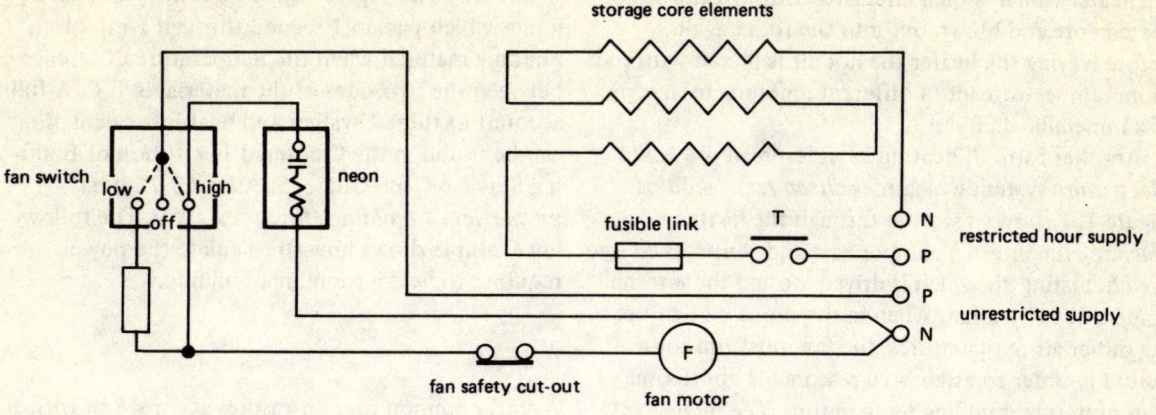

storage core elements

fan switch
low high
off
neon

fusible link   T

N
restricted hour supply
P

P
unrestricted supply
N

fan safety cut-out   F
fan motor

T is a charge control thermostat

Figure 100 *Fan-assisted storage heater*

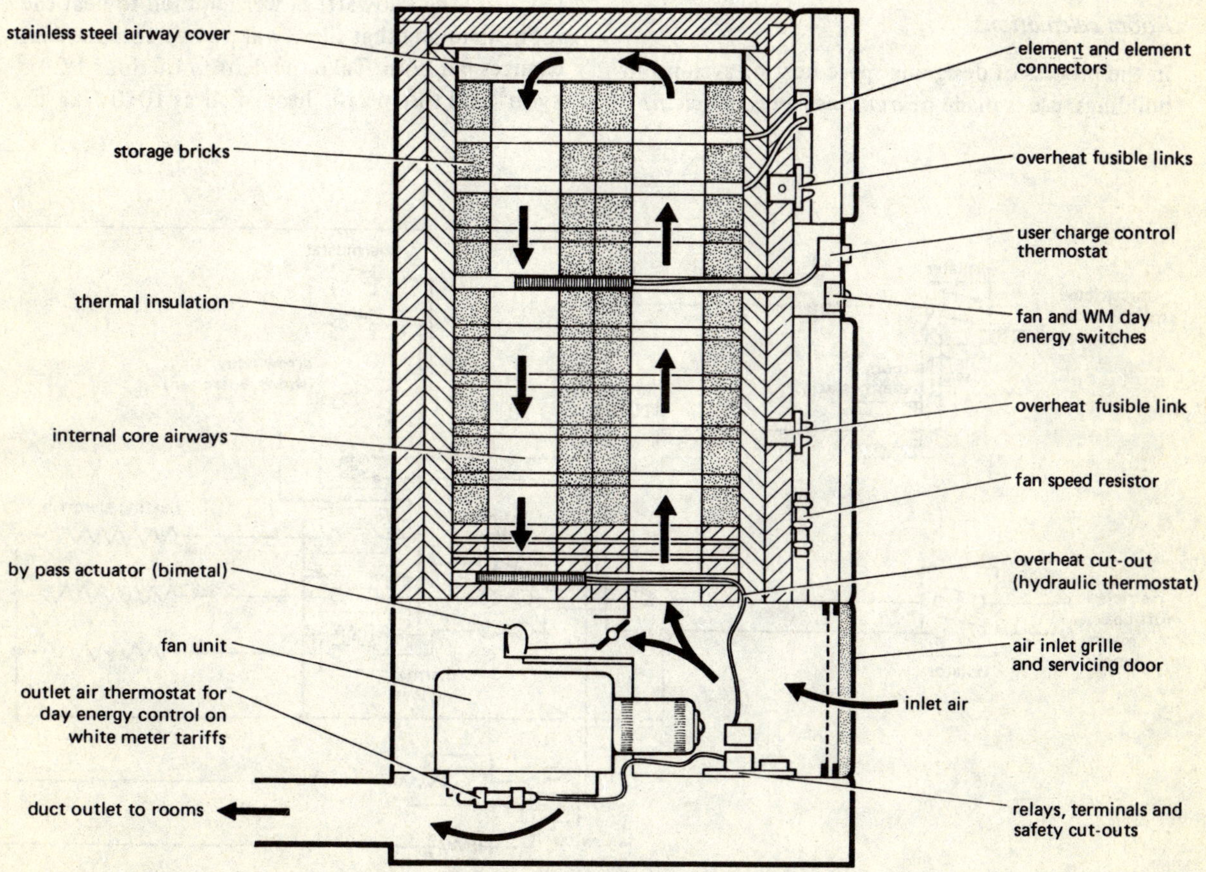

stainless steel airway cover

storage bricks

thermal insulation

internal core airways

by pass actuator (bimetal)

fan unit

outlet air thermostat for
day energy control on
white meter tariffs

duct outlet to rooms

element and element
connectors

overheat fusible links

user charge control
thermostat

fan and WM day
energy switches

overheat fusible link

fan speed resistor

overheat cut-out
(hydraulic thermostat)

air inlet grille
and servicing door

inlet air

relays, terminals and
safety cut-outs

Figure 101 *Section through an Electricaire heater*

the heater which is then circulated through the heated core and blown out into the room again. Before leaving the heater the hot air is mixed with the room temperature air in different amounts by means of a bimetallic damper.

Another form of heating in widespread use is the *Electricaire* system which uses *cheap rate* facilities. Figure 101 shows a section through the heater revealing the internal components and route taken by the circulating air which is driven around the internal core airways by a fan. When in the process of adjusting outlet air temperatures, the fan must run for a period in order to establish a reasonable approximation of normal building temperature. The inlet temperature should be checked to ensure that it is not appreciably below 18°C. Figure 102 is a typical wiring diagram of an Electricaire heater.

Room calculations

In the process of designing space heating systems in buildings, use is made of *transmission coefficients* or

U-values. These represent the quantity of heat in joules which pass in 1 second through 1 m² of building material when the temperature difference between the two sides of the material is 1°C. A full account of these U-values and heat loss calculations can be found in the Chartered Institution of Building Services CIBS Guide, Section A3 *Thermal properties of building structures 1980*. The following example shows how to calculate the power required to heat a room in a building.

Example

A staff common room measures 15 m x 8 m with a ceiling height of 4 m. It has one large window area of 50 m² and one door area of 7 m² and is to be heated electrically so that the internal temperature is maintained at 21°C when the temperature outside is 0°C. Calculate the kilowattt power required to heat the room assuming that there will be two complete air changes per hour. Take the density of air as 1.29 kg/m³ and the specific heat of air as 1010 J/kg°C.

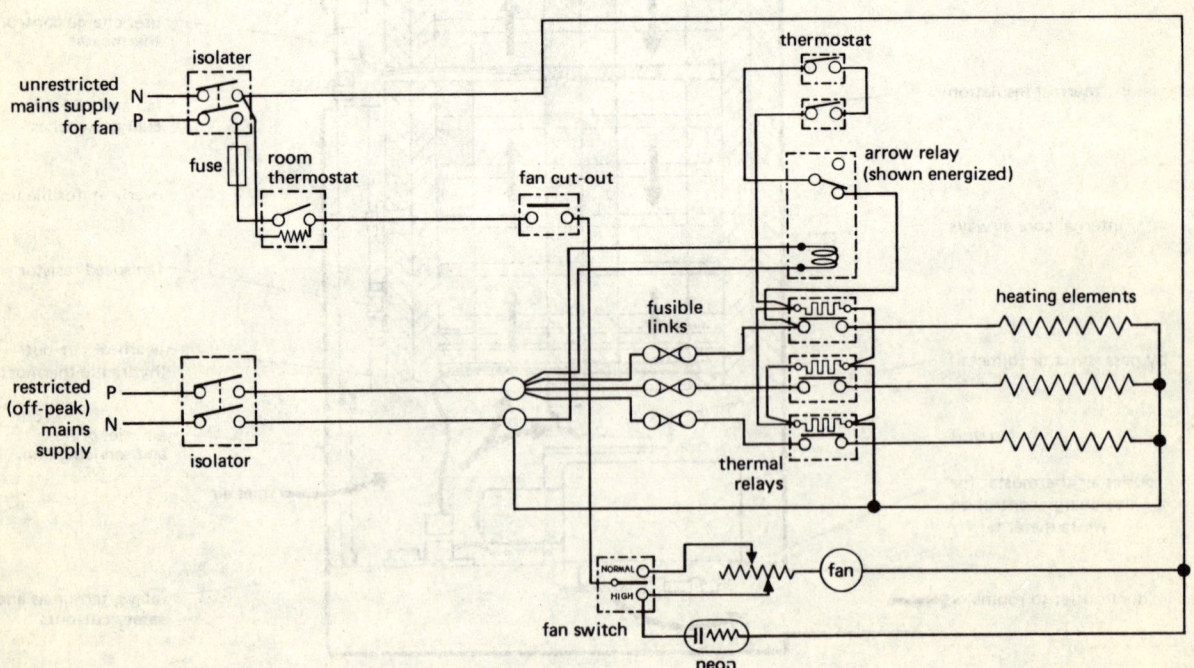

Figure 102   *Wiring diagram of Electricaire heater*

The following U-values can be assumed:

Walls – brick and plastered   1.7 W/m² °C

Floor – timber on concrete   0.8 W/m² °C

Ceiling – plaster   1.2 W/m² °C

Door – wood   2.8 W/m² °C

Window – double glazed   2.9 W/m² °C

The calculation is performed, based on air changes per hour. Firstly, one determines the power required to heat the air; secondly, the power needed to cover the heat losses through the room.

1   Volume of air heated = 2 x 15 x 8 x 4 = 960 m³/hour

Mass of air heated = 960 x 1.29 = 1238.4 kg/hour

Since heat energy required per hour is given by:

$$Q = mc(\theta_2 - \theta_1) \text{ Joules/hour}$$

then $Q = 1238.4 \times 1010 \times 21 = 26.26$ MJ/hour

Since 1 kWh = 3.6 MJ

The energy required to heat the air per hour is

$$\frac{26.26}{3.6} = 7.29 \text{ kWh}$$

The power required is 7.29 kW

2   Heat losses are as follows:

Area of brickwork = area of walls – area of door and window

$$= (2 \times 15 \times 4) + (2 \times 8 \times 4) - (7 + 50)$$

$$= 127 \text{ m}^2$$

Area of floor and ceiling = 15 x 8 = 120 m²

Therefore power loss/°C difference through

walls	= 127 x 1.7 =	215.9 W/°C diff.
ceiling	= 120 x 1.2 =	144.0 W/°C diff.
floor	= 120 x 0.8 =	96.0 W/°C diff.
window	= 50 x 2.9 =	145.0 W/°C diff.
door	= 7 x 2.8 =	19.6 W/°C diff.

Total power loss/°C difference   620.5

The power needed to cover the heat losses is

620.5 x 21 = 13 kW (approx.)

Thus total power required to heat the staff common room = 13 + 7.29 = 20.29 kW

Water heating systems

In terms of these systems, water storage cylinders fed from a cold water supply are the usual practice. The simplest and most popular way of heating the water is by means of immersion heater, either a single heater fitted vertically at the top of the storage cylinder to provide a quick boost of hot water when necessary or a single heater fitted near the bottom to provide large quantities of water. An alternative is to fit two heaters, one at the top and one at the bottom and in this case it is recommended to have a tank size in excess of 137 litres in capacity. One further arrangement quite commonly used is to fit a dual-element heater as shown in Figure 103. Each of these is normally controlled by its own thermostat with the longest element being supplied on Economy 7 tariff.

Other methods of water heating include cistern-type water heaters such as shown in Figure 104 and point-of-use heaters such as the non-pressure type with its cold water inlet and swivel hot water outlet, see Figure 105. The cistern type is used for bulk storage of heated water for supplies to numerous hot taps whereas the point-of-use heaters are used over basins and sinks and sometimes baths for an immediate supply of hot water. Also available are instantaneous heaters with the water passing through a small heating unit which has an in-built pressure switch to prevent the element being switched on before the water starts to flow, and instantaneous shower heaters with built-in safety features to prevent overheating. Figure 106 shows a typical instantaneous unit.

Figure 107 is a typical modern heating system using free solar energy.

With regard to corrosion, heating systems in hard water areas are prone to sludge or 'black rust'. This sludge is about five times heavier than water and forms a dense sediment which spreads from the bottom of radiators. Once scale starts to clog the system, thermostat settings have to be higher to maintain the required heating level. If left untreated, the heating system will become less efficient, valves will

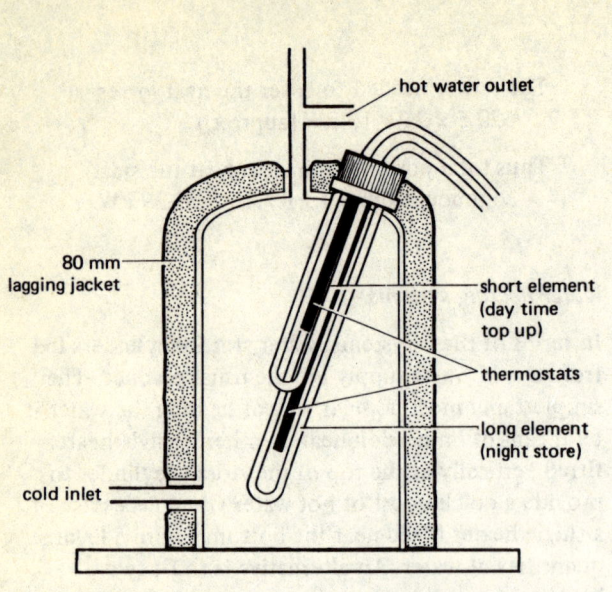

Figure 103 *Dual-element immersion heater*

hot water outlet

80 mm
lagging jacket

short element
(day time
top up)

thermostats

long element
(night store)

cold inlet

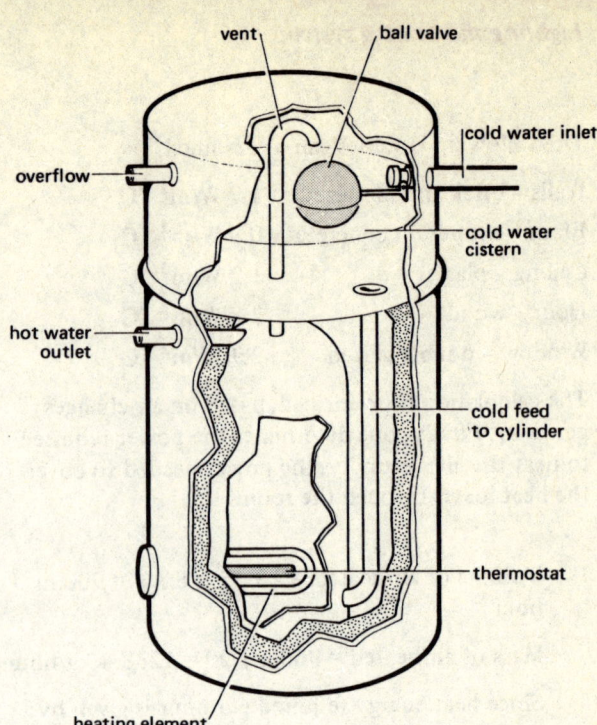

Figure 104 *Cistern-type water heater*

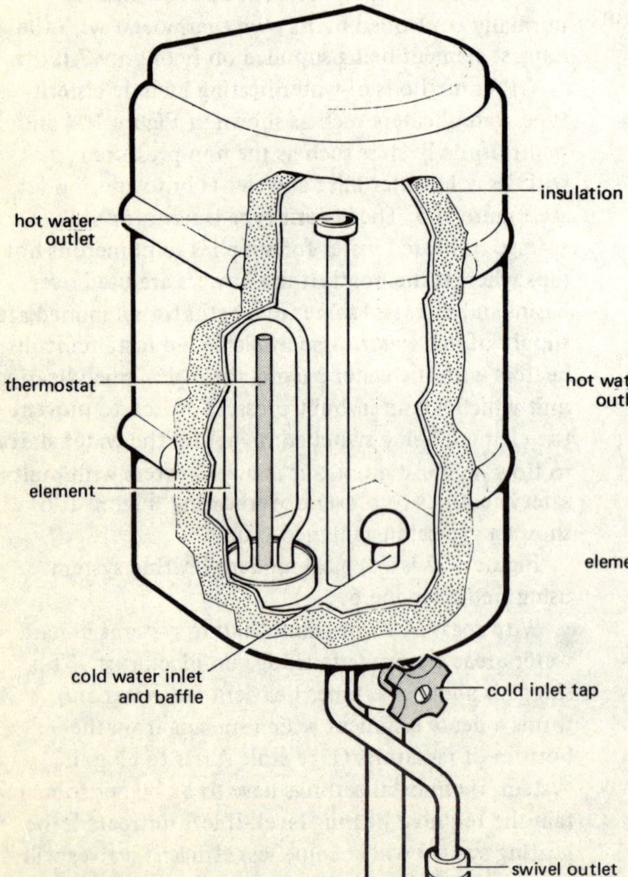

vent

ball valve

cold water inlet

overflow

cold water
cistern

hot water
outlet

cold feed
to cylinder

thermostat

heating element

insulation

hot water
outlet

thermostat

element

cold water inlet
and baffle

cold inlet tap

swivel outlet

Figure 105 *Non-pressure water heater*

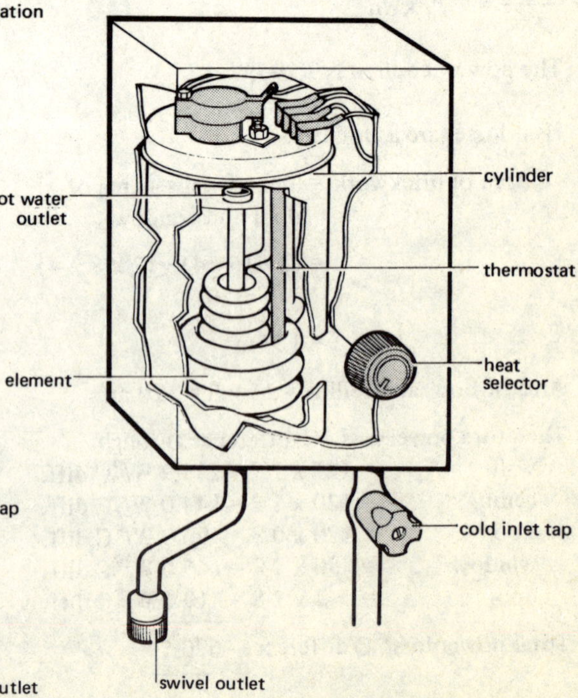

cylinder

hot water
outlet

thermostat

heat
selector

element

cold inlet tap

swivel outlet

Figure 106 *Instantaneous water heater*

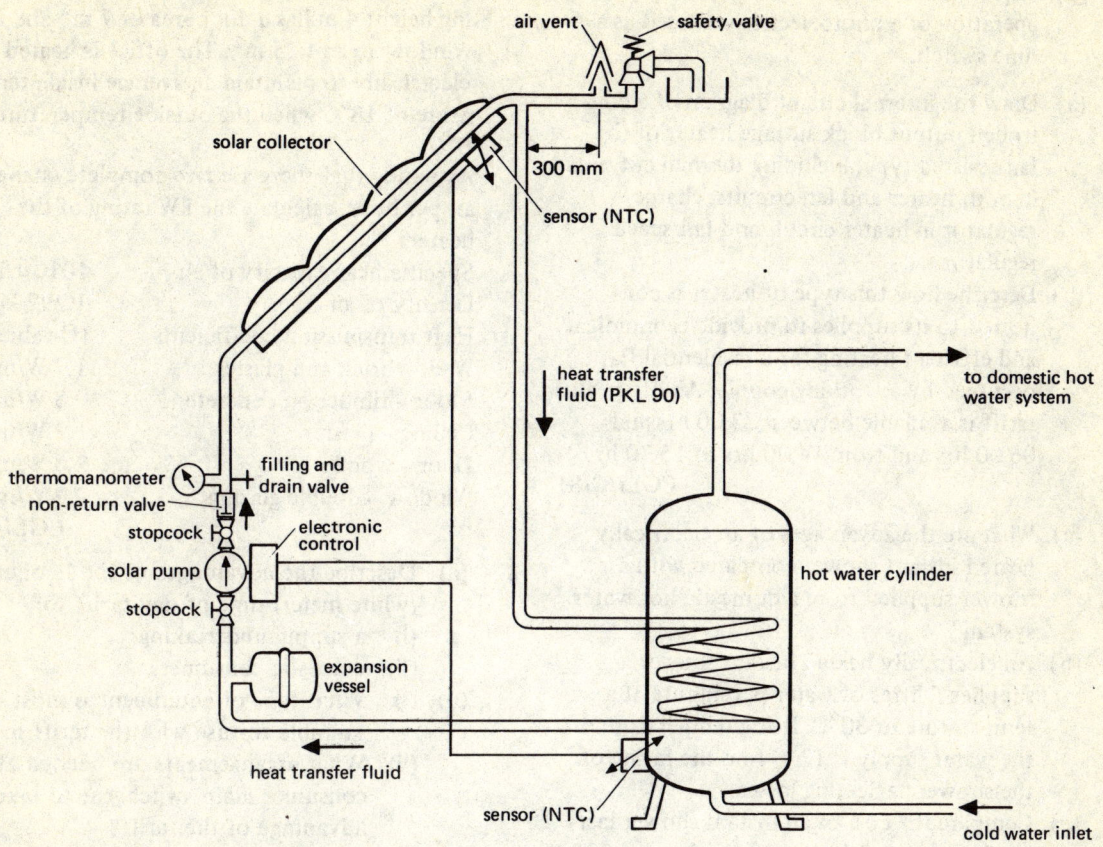

air vent

safety valve

solar collector

300 mm

sensor (NTC)

heat transfer
fluid (PKL 90)

to domestic hot
water system

thermomanometer

filling and
drain valve

non-return valve

stopcock

electronic
control

solar pump

hot water cylinder

stopcock

expansion
vessel

heat transfer fluid

sensor (NTC)

cold water inlet

Figure 107 *Solar system for heating water*

stick, pumps will not work, knocking sounds will be heard and pipes will become furred up. The system will eventually become very expensive to rectify.

Revision exercise 4

1 Explain, with the aid of a diagram, the operation of a high pressure sodium vapour discharge lamp. State the lamp's colour rendering, efficacy, normal position and application.

2 A neon sign is to be installed on the fascia of a shop front. Explain with the aid of sketches, the method of
   (a) terminating the HV metal sheathed cable at the neon tube

   (b) terminating the HV metal sheathed cable at the transformer casing
   (c) fixing the discharge tubes to their metal letters

CGLI/C/77

3 (a) List *five* points to be considered when dealing with the cleaning and maintenance of high-bay luminaires.
   (b) What precautions ought to be considered when designing and installing lighting in a hospital ward area?

4 (a) What precautions are needed when disposing of discharge lamps, particularly sodium and mercury types?

(b) Explain with the aid of a circuit diagram, the operation of a photoelectric cell used as a time switch.

5 (a) Draw the internal circuit diagram of a controlled output block storage heater of the fan assisted type, including thermal cut outs in both heater and fan circuits, charge regulator in heater circuit and fan speed regulator.

(b) Describe how this type of heater is connected to its supplies to provide economical and efficient heating for a residential flat occupied by a working couple. An off-peak tariff is available between 23.00 hrs and 06.00 hrs and from 14.00 hrs to 15.30 hrs.

CGLI/C/81

6 (a) What are the advantages of an electrically heated instant shower compared with a shower supplied from a domestic hot water system?

(b) An electrically heated instant shower supplies 5 litres of water per minute at a temperature of $30°C$. If the temperature of the water supply is $12°C$ find the rating of the shower neglecting losses.

(c) Compare the cost of an instant shower lasting 3 minutes with that of a bath using 135 litres of water at the same temperature supplied from an electrically heated hot water system having an efficiency of 95 per cent. The cost per unit is 5p. The specific heat capacity of water is $4.2 kJ/kg°C$.

CGLI/C/82

7 A works reception office 12 m x 10 m with ceiling height 4 m has a door area of $7 m^2$ and a window area of $25 m^2$. The office is heated electrically to maintain an average inside temperature of $18°C$ when the outside temperature is $0°C$.

Assuming that there are *two* complete changes of air per hour, calculate the kW rating of the heaters.

Specific heat capacity of air	1010 J/kg°C
Density of air	1.292 kg/m³
Heat transmission coefficients	(U-values)
Walls – brick and plaster	1.7 W/m² °C
Floor – timber on concrete	0.8 W/m² °C
Ceiling – plaster	1.2 W/m² °C
Door – wood	2.8 W/m² °C
Window – double glazing	2.9 W/m² °C

CGLI/C/80

8 (a) Describe the advantages of the Economy 7 (white meter) time of day tariff to:
    (i) a supply undertaking
    (ii) domestic consumers

(b) (i) What type of equipment is most suitable for use with the tariff in (a)?
    (ii) What arrangements are needed at the consumer main switchgear to take full advantage of the tariff?

(c) Draw a simple diagram showing how a 40 A special circuit may be fed from the high rate tariff when the weather conditions merit higher cost energy consumption. Assume that the arrangements described in (b)(ii) exist.

CGLI/C/84

chapter five

Planning and estimating

After reading this chapter you will be able to:

1. Assess the constraints surrounding large projects and the need to have project co-ordination.
2. Understand the basic principles of network analysis.
3. Draw networks from simple projects and derive bar charts associated with work activities.
4. Understand the site duties of an electrical sub-contractor.
5. Know the basic elements of estimating.

Planning

A factor which has emerged with the use of sophisticated planning techniques is an awareness by contractors of the full effects of disruptions on construction programmes, often caused by the late issuing of information by consultants. As a result, contractors are today more mangement orientated, or better still, system conscious people, capable of presenting programmes and claim reports in an effective and concise manner. This relatively fast transformation has caused considerable dissent in the industry between contractors and consultants who themselves have seen considerable changes in the industry over the last few years in which building work has become a highly fragmented operation demanding the co-ordination by the contractors of many different specialist suppliers and sub-contractors. A relationship chart of the industry is shown in Figure 108.

It is generally not unfair to say that consultants and local authorities, with a few notable exceptions, have not developed their knowledge of managerial techniques to meet the increasingly demanding requirements of the industry as it now exists. The resulting situation is one where contractors are attempting to apply a highly planned approach to their projects when many of the possible benefits have already been obviated by the shortcomings of consultants who are unable to co-ordinate their own activities in anticipation of the probable requirements of the construction team. As a result an effective remedy is to provide some form of *project co-ordination*.

Project co-ordination, as it implies, should ideally be instigated at the inception of a major contract or scheme, rather than at the time when the contractor takes possession of the site. Too familiar is the often inevitable confrontation at initial site meetings of consultant and contractor whereby the contractor demands the immediate issue of various categories of unavailable information which he claims to be vital to the progress of his construction programme. From the very offset of the construction work, a divisive influence is experienced as a result of bad communication which could have been eliminated by informed forward planning.

A few contracting companies, notably those working on a fee basis, have sought to eliminate any conflict by taking responsibility for the project management from the beginning and this has been an acceptable compromise to some clients, but unfortunately it does tend to be viewed with some reservations as the contractor is required to act in the interest of the client while simultaneously protecting the eventual profitability of his own operations. There is a belief in the industry that project co-ordination should be carried out by consultants and commenced at the very beginning of a scheme.

By using an independent planning consultant an optimum of control can be achieved since no bias is being shown towards any particular function within the overall operation. The client is able to receive a totally independent and impartial service to assist him in the control of his project. This introduction of management control will inevitably help to ensure that the most economic and effective work flow is

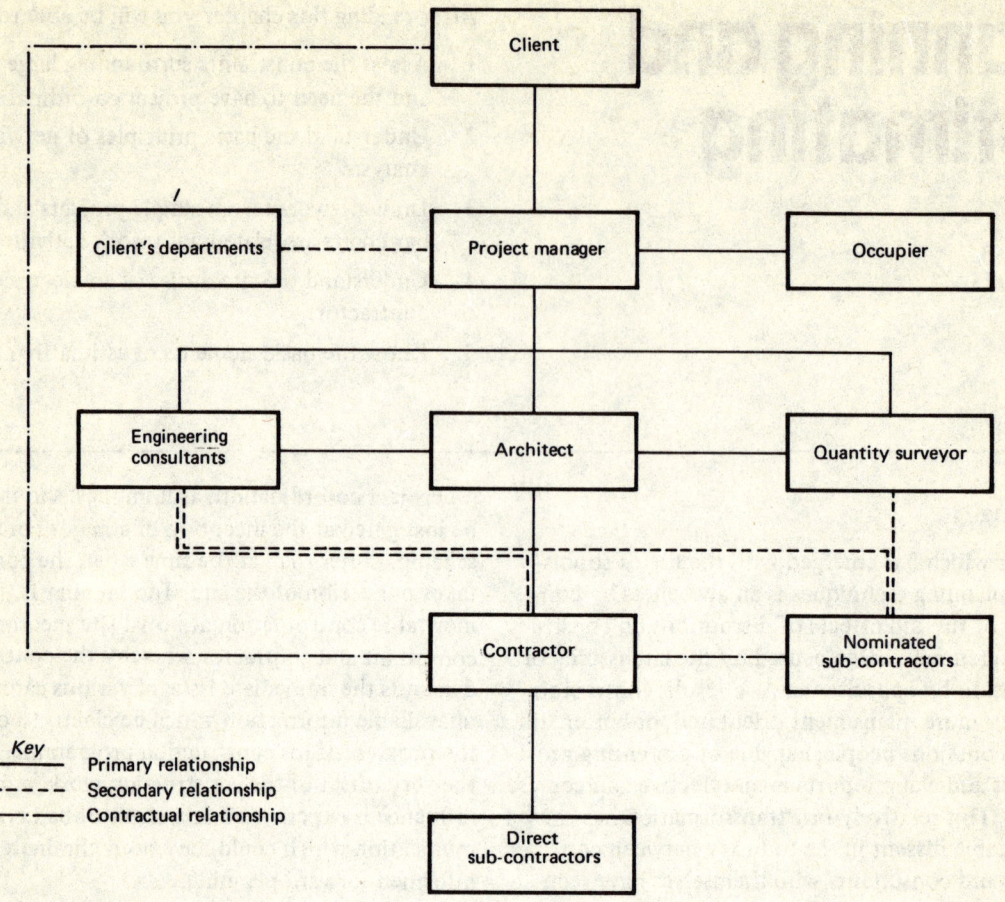

Figure 108 *Large project relationship chart*

achieved by both the client's staff and the consultants in order that the most favourable progression of the construction programme is permitted from the time that the contractor establishes himself on site. The contractor's performance can be monitored and accurately assessed, and tangible benefits obtained from the elimination of unjustified claims for extra costs and time.

Project co-ordination

Basically, this means the forward planning of a project or whole series of projects, and it is done in advance of the start of pre-contract design work using modern programming techniques such as network analysis. It is seen as the control of a project through all its stages by a system of regular review which provides the current picture at all times and allows frequent comparisons between actual progress and targets for that time. Integral in the system of planning and control is the allocation of resources, both financial and manpower. It is by critical path network programming that it is possible to see in advance and make provisions for staff needs at any stage of the project, as well as detect any sudden increase in cost and take the necessary action to remedy the matter.

Project co-ordination, then, enables clients to plan with confidence for site handover dates and to achieve better communications and efficient use of staff and resources. This will ultimately lead to faster and cheaper building and to earlier returns on capital.

The procedure involving project co-ordination is by taking the earliest operations in the programme and assigning them to the client's staff and consultants. Each individual or consultants' representative is asked to give a time estimate for each of the tasks or activities in which he or the staff of his department are involved. Estimates may be based on previous experience, or, if similar experience is lacking, be based on a carefully considered estimate. The various activities are then formed into a logical order on a diagram that will form the basis of a network for the project.

The work of producing the original project pro- gramme, updating programmes and schedules, monitoring the progress of individual activities and keeping members of the project team constantly informed of the current situation is a complex and time consuming part of the system. To ensure that this important work is adequately carried out, it is advisable to appoint an individual to be responsible for it and who has had the necessary specialized train- ing in the use of modern planning techniques. This person will be appointed *project co-ordinator* and although he does not hold executive responsibility, he will have the responsible job of placing all available information before the *project manager* and for pointing out the consequences of alternative courses of action. It is the project manager's job to see that the project is organized properly, to set up a team who will undertake the work, prepare the plan of action and monitor progress as well as act as a link during all of the project stages. This will involve such things as dealing with the customer or client, writing reports, chairing meetings and watching cash flow and profit situations.

Network analysis

For large development projects, network analysis has emerged as one of the best methods of planning and control. It brings the project programme into the office of every individual concerned and it auto- matically relates one's role to that of colleagues and other individuals in other departments of work. Contrary to common belief, network analysis is simple, easily explained and understood and offers a number of advantages. The very preparation of a net- work, in whatever form it takes, forces the project team to think deeply and carefully about the project. It demands the early logical planning of the whole project and shows clearly how all the operations within the project are interrelated. It can also be used to find the most economical use of time and method of carrying out the project considering the available resources and it controls and relays progress of the project, providing an instant picture of the current situation. Moreover, it becomes an extremely quick and flexible method of revising plans when delays occur, and because it is purely a statement of logic which remains constant whether activities are longer or shorter in time than estimated, the network itself does not become redundant. It ensures that the responsibility for each activity is clearly defined and each person engaged within it sees when he should complete his work as well as see what others are doing.

To produce a network, the project is broken down into possibly a large number of jobs which are often described as operations or activities. To do this the project co-ordinator has to obtain as much informa- tion as possible on everything contained within the project. He has then to list the information in a formal manner such that the relationships between each activity and those which precede it and follow it are logically presented. This will involve such things as the estimated time that a certain activity will take as well as dates before which it can- not start or must finish. Once the list has been drawn up, calculations are done to show the period of time each activity is likely to take. There are two dates given to this period of time, namely, *earliest start date* and *latest finish date*.

There are several approaches to drawing networks, but the two most widely used are called Arrow Diagram and Precedence Method. In the Arrow Diagram, arrows drawn with solid lines represent activities involving a period of time which is inserted beneath the line while above the line is inserted a brief description of the activity. Arrows drawn with broken lines are called *dummies* and represent logical restrictions upon succeeding activities. They are essential for the construction of the network and indicate the dependency of the activities upon each other: they have no time allocation. There is no significance given to the length of lines but they will be joined at both ends by circles (and sometimes other shapes) called *events* or *nodes*. It will be seen that the tail of an arrow indicates the beginning of an activity while the head indicates the finish. Normal practice is to work from left to right, starting the project with event one and moving through to the completion with a final event number. Events can either be drawn into three sections or quadrants so as to provide the network with information about the earliest event time and latest event time – the former being inserted on the left-hand side of the event and the latter

inserted on the right-hand side. In cases where more than one arrow enters an event, the practice is to insert a time unit from those activities taking the longest time to complete. One moves through the network this way, inserting, first, all the earliest event times (see Figure 109).

The network is assessed so as to determine the minimum time which must elapse before the project can be completed. The longest route through the network to meet this requirement is called the *critical*

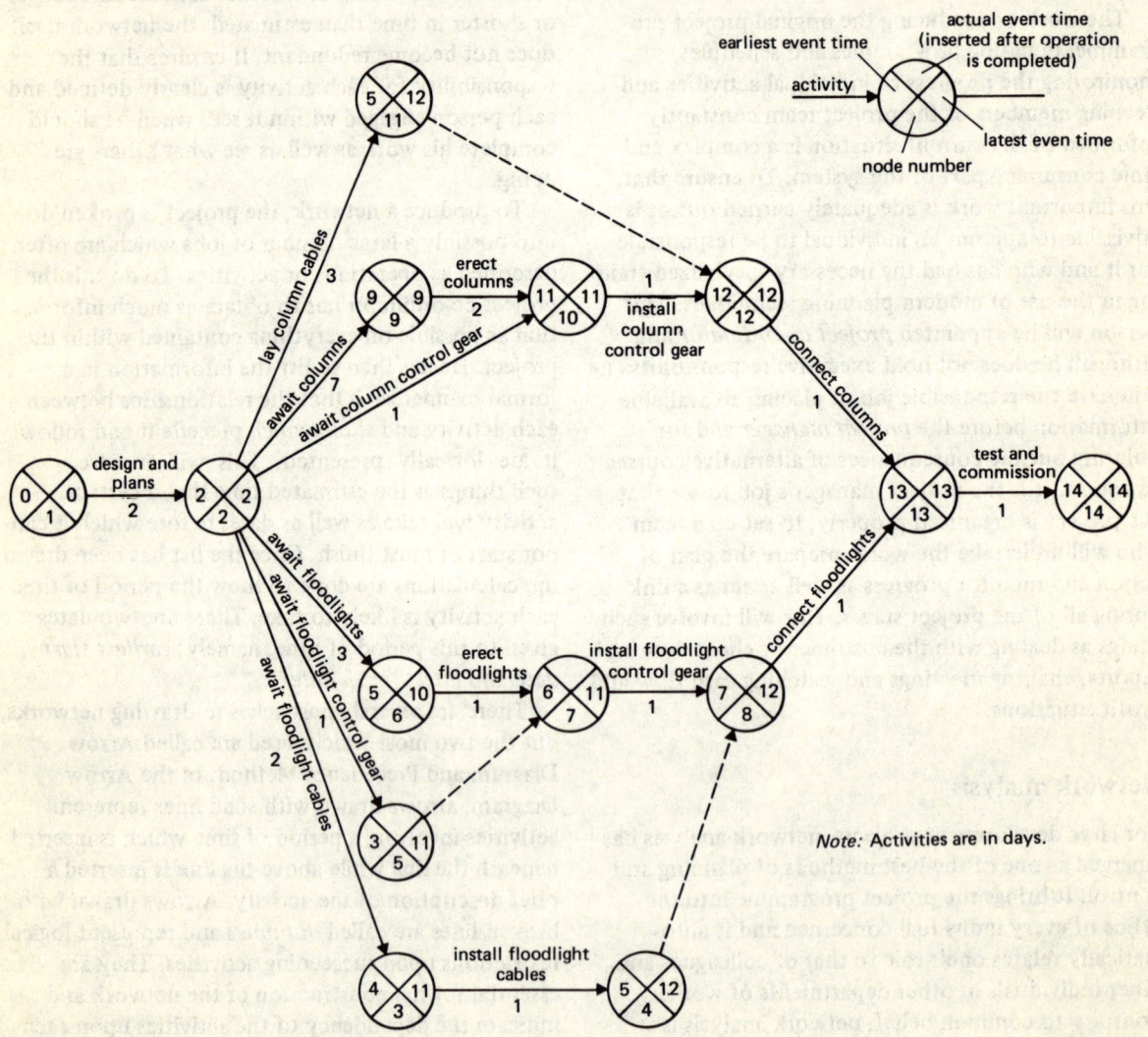

Note: Activities are in days.

Figure 109   *Simple CPA network for installing floodlights*

Activities	Time (days)	Total float (days)		Critical Path
1 − 2	2	None	0	CP
2 − 3	2	11-2-3	= 6	
2 − 5	1	11-2-1	= 9	
2 − 6	3	10-2-3	= 5	
2 − 9	7	None	0	CP
2 − 10	1	11-2-1	= 8	
2 − 11	3	12-2-3	= 7	
3 − 4	1	12-4-1	= 7	
6 − 7	1	11-5-1	= 5	
7 − 8	1	12-6-1	= 5	
8 − 13	1	13-7-1	= 5	
9 − 10	2	None	0	CP
10 − 12	1	None	0	CP
12 − 13	1	None	0	CP
13 − 14	1	None	0	CP

Figure 110   *CPA of floodlighting project*

path. An analysis of non-critical activities is done to determine to what extent activities may be delayed without affecting the overall project time and the start of subsequent activities. By working back through the network, the non-critical activity times are subtracted from succeeding event latest times starting with the final project time. This method is

illustrated in Figure 110 and it will reveal from the network any spare time that exists. This spare time is called *float* and is the greatest difference between earliest and latest event times. Only a simple network has been shown and Figure 111 is a bar chart.

In the Precedence Method, the activities are shown as boxes and the relationship between them is again through connecting arrows. These arrows, however, do not indicate the length of activity time. The network is read such that an activity should not be started until all the immediately preceding activities have been finished. Inserted on each box activity, reading from left to right, will be the information regarding earliest start time, duration of the activity, latest finish time and float. Normally it is the practice to write a brief description inside the box of what the activity involves. The advantage of this method is in the positioning of the activities in time since it allows the project manager to quickly estimate such things as cash requirements and resource requirements.

Project management procedures

In order to provide some 'insight' into project management procedures, one first needs to have an understanding of the various stages likely to be encountered in any given project. These stages are set out as shown

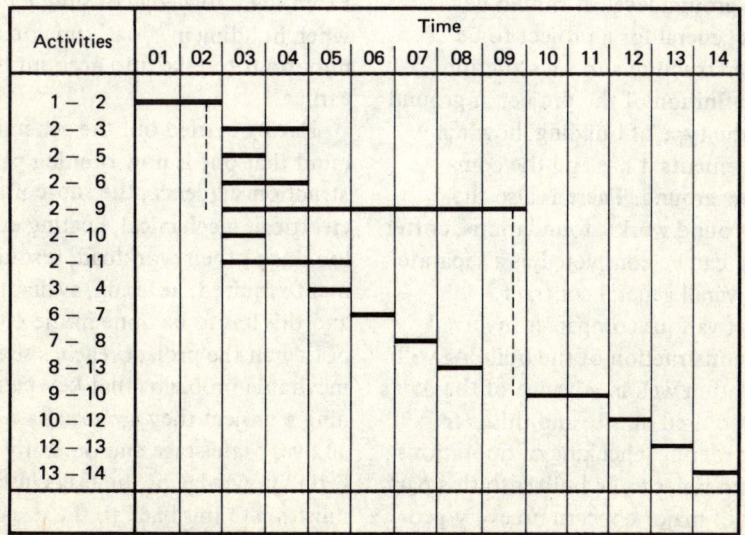

Figure 111   *Bar chart for floodlighting project*

in Figure 112. In practice, the majority of tasks undertaken by design consultants fall within stages 1, 2 and 3. This brings the project through tender and tender analysis stage and allows the professional companies involved to report to their clients on the contractors and their ability to carry out the project as designed.

On reaching stage 4, the project will already have its general contractor and possibly nominated sub-contractors and suppliers and a series of quotations and information which have arisen out of the enquiries made by the design consultants, architect, structural engineer and mechanical and electrical engineers. Such information will be subsequently passed on to the successful contractor.

With regard to the mechanical and electrical consultants or other specialists which are allied to the heating, ventilating and air-conditioning systems, the design can be developed a stage further where necessary, showing how the systems fit into the building, their relationship to the structure and other design features, and in parallel with this, quotations and estimates are sought from other specialists which will allow the engineering services to function, for example, control systems, pumps, boiler plant, air-handling equipment etc.

At this stage the general contractor would have started on site and will be calling for any of the building services which need to go into the substructure or work which is below ground level. It should be pointed out that it is general for a project to be divided between superstructure and substructure which gives a clear definition of the project at ground level, depending on the type of building, how many basements or sub-basements it has and the complexity of work below ground. There is also the possibility that the ground works, foundations, coffer dams, piled walls etc. can be completed as a separate pre-contract to the overall general contract.

It is inevitable that various companies involved with the design and construction of the building will be calling for information well in advance of the dates when equipment is required on site and this is to allow pre-planning, ordering, checking of quotations and for manufacturing times to be built into this particular sequence. It is of major concern on every project and it is one that is often badly managed. There cannot be a situation on a construction project when

one could use the expression that 'it will be alright in the end' because most of the problems that arise during the progress of the work relate to the lack of thought that goes into the analysis of information and also the design and detailing of each element relative to the overall project itself. The industry has been trying to improve upon these particular problems over a number of years. It is therefore important to understand the contractual relationship between the companies involved and how individual persons perform before one can appreciate why some problems manifest themselves.

The correct sequence to be employed when programming a project is to systematically analyse each element. This should cover the tasks to be performed from design through to manufacture and final installation of the product concerned, for example, ductwork control systems, lighting equipment etc. The complete sequence of events can then be plotted from the date when the enquiry is sent out by the designer or company concerned to completion. This can be prepared in chart form with an estimate of time included for human error and also for the passing of information between various parties. This is fine for one element, but when you bring together a number of interrelated control systems then it is important to ensure that a network is prepared showing the interrelationship between these functions and where there is a dependency of one upon another. From the knowledge of simple network analysis, when building in 'float' time on a network it becomes important to take into account the points mentioned earlier.

Having carried out the general planning, bearing in mind that one is now running parallel with the construction sequence, the sub-contractors (for example, electrical, mechanical, heating etc.) are obviously looking at their workload, resources, plant and equipment required, ordering, availability of materials etc. and this has to be done in the company's office and not when the project reaches site. This is one of the inevitable problems that key people have. When running a project they are usually brought into the work at a very late stage and on many occasions have very little knowledge of the tasks in front of them. One must relate this back to the designers, architects and others who have been involved possibly for a year, maybe 18 months, before the events take place on

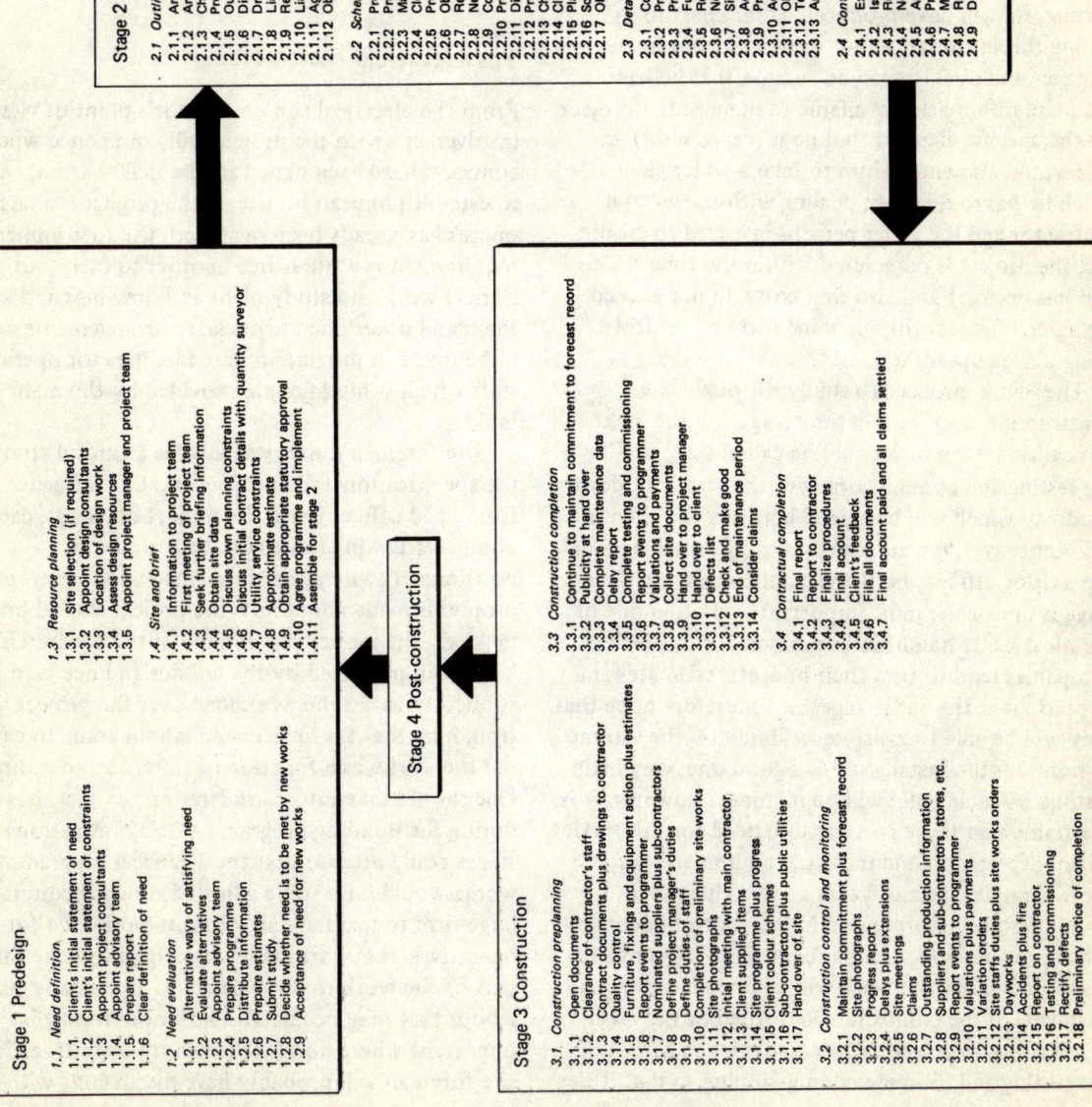

Stage 1 Predesign

1.1 Need definition
1.1.1 Client's initial statement of need
1.1.2 Client's initial statement of constraints
1.1.3 Appoint project consultants
1.1.4 Appoint advisory team
1.1.5 Prepare report
1.1.6 Clear definition of need

1.2 Need evaluation
1.2.1 Alternative ways of satisfying need
1.2.2 Evaluate alternatives
1.2.3 Appoint advisory team
1.2.4 Prepare programme
1.2.5 Distribute information
1.2.6 Prepare estimates
1.2.7 Submit study
1.2.8 Decide whether need is to be met by new works
1.2.9 Acceptance of need for new works

1.3 Resource planning
1.3.1 Site selection (if required)
1.3.2 Appoint design consultants
1.3.3 Location of design work
1.3.4 Assess design resources
1.3.5 Appoint project manager and project team

1.4 Site and brief
1.4.1 Information to project team
1.4.2 First meeting of project team
1.4.3 Seek further briefing information
1.4.4 Obtain site data
1.4.5 Discuss town planning constraints
1.4.6 Discuss initial contract details with quantity surveyor
1.4.7 Utility service constraints
1.4.8 Approximate estimate
1.4.9 Obtain appropriate statutory approval
1.4.10 Agree programme and implement
1.4.11 Assemble for stage 2

Stage 2 Design

2.1 Outline design
2.1.1 Analyse brief
2.1.2 Analyse site information
2.1.3 Check estimate of cost
2.1.4 Prepare outline design
2.1.5 Outline planning approval
2.1.6 Discuss contract proposals with quantity surveyor
2.1.7 Draft outline specification
2.1.8 Liaise with interior designer
2.1.9 Report event to client
2.1.10 Liaise with local authority and public utilities
2.1.11 Agree design and obtain client's approval
2.1.12 Obtain financial approval

2.2 Scheme design
2.2.1 Preliminary project budget
2.2.2 Prepare final sketch design
2.2.3 Maintain liaison with local authority on planning matters
2.2.4 Clear exceptional contract procedures with quantity surveyor
2.2.5 Prepare outline specifications
2.2.6 Obtain estimate, etc. as appropriate
2.2.7 Report events to client
2.2.8 Negotiate with public utilities
2.2.9 Consult fire officer
2.2.10 Project manager on sketch design
2.2.11 Discuss wayleaves, etc. with property department
2.2.12 Procurement
2.2.13 Check compliance with building regulations
2.2.14 Client approval of design
2.2.15 Planning acceptance
2.2.16 Schedule fuel requirements
2.2.17 Obtain financial approval

2.3 Detail design
2.3.1 Cost check and final budget project
2.3.2 Production and tender information
2.3.3 Pre-tender action for main contract
2.3.4 Furniture, fixtures and equipment revised estimate
2.3.5 Report events to programmer
2.3.6 Nominated suppliers plus sub-contractors
2.3.7 Site control staff
2.3.8 Advance ordering
2.3.9 Preliminary site works
2.3.10 Advance publicity
2.3.11 Obtain authorization
2.3.12 Tender data
2.3.13 Approval to issue tenders

2.4 Contract preparation
2.4.1 Examine lowest tender plus prepare reconciliation statement
2.4.2 Issue tenders
2.4.3 Report events to programmer
2.4.4 Nominated suppliers plus sub-contractors
2.4.5 Advance ordering
2.4.6 Preliminary site works
2.4.7 Main contract documentation
2.4.8 Report to client
2.4.9 Discuss site works with architect

Stage 4 Post-construction

Stage 3 Construction

3.1 Construction preplanning
3.1.1 Open documents
3.1.2 Clearance of contractor's staff
3.1.3 Contract documents plus drawings to contractor
3.1.4 Quality control
3.1.5 Furniture, fixings and equipment action plus estimates
3.1.6 Report events to programmer
3.1.7 Nominated suppliers plus sub-contractors
3.1.8 Define project manager's duties
3.1.9 Define duties of staff
3.1.10 Completion of preliminary site works
3.1.11 Site photographs
3.1.12 Initial meeting with main contractor
3.1.13 Client supplied items
3.1.14 Site colour schemes
3.1.15 Sub-contractors plus public utilities
3.1.16 Sub-contractors plus public utilities
3.1.17 Hand over of site

3.2 Construction control and monitoring
3.2.1 Maintain commitment plus forecast record
3.2.2 Site photographs
3.2.3 Progress report
3.2.4 Delays plus extensions
3.2.5 Site meetings
3.2.6 Claims
3.2.7 Outstanding production information
3.2.8 Suppliers and sub-contractors, stores, etc.
3.2.9 Report events to programmer
3.2.10 Valuations plus payments
3.2.11 Variation orders
3.2.12 Site staffs duties plus site works orders
3.2.13 Dayworks
3.2.14 Accidents plus fire
3.2.15 Report on contractor
3.2.16 Testing and commissioning
3.2.17 Rectify defects
3.2.18 Preliminary notice of completion

3.3 Construction completion
3.3.1 Continue to maintain commitment to forecast record
3.3.2 Publicity at hand over
3.3.3 Complete maintenance data
3.3.4 Delay report
3.3.5 Complete testing and commissioning
3.3.6 Report events to programmer
3.3.7 Valuations and payments
3.3.8 Collect site documents
3.3.9 Hand over to project manager
3.3.10 Hand over to client
3.3.11 Defects list
3.3.12 Check and make good
3.3.13 End of maintenance period
3.3.14 Consider claims

3.4 Contractual completion
3.4.1 Final report
3.4.2 Report to contractor
3.4.3 Finalize procedures
3.4.4 Final account
3.4.5 Client's feedback
3.4.6 File all documents
3.4.7 Final account paid and all claims settled

Figure 112 Plan of work for project management procedures

site, and, therefore, do have a detailed knowledge of the reasons why the design has been tackled in a particular way.

If the sub-contractors are competent companies, which one must assume they are, they would have also looked at the value of the work, the time scale and the amount of labour, site supervisors, senior staff, technicians etc. that will be required to complete the project in the time scale that has been set. This makes due allowance for time which is lost due to variations, changes, disruption, holidays, sick leave etc., and a company that is resourceful will have produced a 'bar chart form' setting this information out so that the contracts manager and director can be assured that he has adequate information available to manage the project for the specific element that he is responsible for. This would also enable him to take a wider view, which he has to do when dealing with the general contractor and the other persons involved to ensure that the project is completed within the time-scale that has been set and also that costs do not exceed the agreed figures without some form of control being placed upon them.

The work proceeds rapidly through the construction and installation stage to the next critical element which could be called stage 5. This is the testing and commissioning of the services prior to handover which will be agreed between the client's representatives, the architectural project manager or supervising officer and the general contractor. This stage is one of the most important and often one of the most badly handled. The reason for this is that companies tend to trim their budgets, estimates and quotations at the early stages and therefore hope that they will be able to satisfy everybody on the assumption that, if the installation is a good one, very little testing, balancing etc. will be required. However, it is inevitable that there is even more need for this service when information and technical equipment become more complex as the years go by. These tasks should also be incorporated in the control documentation, the network, the bar charts and labour histograms and either the work will be carried out by the engineers of the companies concerned or they will invite offers from specialist companies to carry out the testing and ultimate commissioning so that it ties in with the performance criteria that was set at the beginning of the contract by the consulting engineers or the client.

When one comes to the final handing over and the balancing of the engineering systems to see if they are working and properly installed, the client will be in occupation. Problems often arise and the design team and the contractors spend a great deal of time arguing over what is wrong and who is responsible. The only way to obviate these problems is to try and allow for a form of technical handover which will allow the majority of work to be completed before the official handover date.

The electrical contractor

From the electrical sub-contractor's point of view, his involvement with the project will commence when contracts have been signed and he makes arrangements to establish himself on site. If the project is a large one, as has already been suggested, the first immediate requirement is a site office in order to carry out clerical work and study plans and drawings and keep these and other documents safe. Arrangements need to be made to provide welfare facilities for operative staff which sometimes are provided by the main builder.

Site foreman duties will involve a careful study of the specification and any notes that are attached to it from head office. From the project drawings, cable/conduit/trunking routes can be sorted out, positions of switchgear ascertained and numerous proposals made about how the project should be tackled. The master programme of the works will have been produced by the builder and needs to be studied to assess the work load over the project. It is from here that the anticipated labour force to carry out the works, as a function of time, can be estimated. One should take into consideration the fact that during the building programme delays at various stages could arise so that the electrical programme of works would have to be adjusted. For a medium to large firm to maintain an efficient work load for its operatives, it is worthwhile investigating some other part of the work to be available to absorb any surplus labour that may occur. This latter point may be important where bonus schemes are operating. The site foreman will probably have discussions with the electricity supply authority with regard to intake points and cable runs and also discussions with the main contractor for temporary supplies to the site. Using the labour force available to him, he must now work out on what sections of the project he should deploy his men.

As the project progresses, there will be a need to have regular site meetings with the main builder and other trades to discuss difficulties where and when they arise. It is at these meetings that variations of the work will be discussed and AIs issued covering such variations. It is important that the site foreman informs head office of any changes. Verbal requests need to be passed on specifying details of what has to be done in terms of materials and labour required and when the changes need to be made and need to be completed. On a well run site the builder usually takes minutes of the meetings and any points not to the satisfaction of the site foreman should be cleared before the minutes are signed.

With regard to movement of labour, unnecessary movement of operative staff is most inadvisable because this makes responsibility for their tasks seem somewhat pointless, particularly if they are unable to see its completion. Where movement of labour is essential it is good management/labour relations if the same operatives return to the project, pursuing a different phase of their original work.

Estimating

Estimating is a system of compiling information to facilitate competitive tendering and is seen as the technical process of predicting net or prime costs. It will involve the studying of contract documents, drawings and specification etc. in order to assess the material and labour necessary for carrying out the work.

The initial steps start off with an 'enquiry' or invitation to tender. For example,

Dear Sir,

Area Transmission Office – Whitchurch Rd, Cardiff

We write to invite your tender for the electrical sub-contract to the above scheme. To enable you to prepare your tender, we enclose the following:
1   One copy of each drawing – Nos. AM 139–45.
2   Electrical specification.

3   Standard form of tender for nominated sub-contractors, pages 1–6. The tender *will not* be considered unless submitted on this form.
4   RIBA form of agreement between employer and nominated sub-contractor 1971.
5   Daywork schedule.
6   Tender summary.
7   Return envelope.
Additional notes.

These additional notes will be important since along with other things they will stipulate the basis of the tender by indicating whether the estimate is to be a fixed price quotation or one that is based on a fluctuating price, that is, one which will need to be adjusted for changes in costs submitted at a later date. For example, the additional notes might read:

(a)   Alternative tenders are required. Tenders on a fluctuating basis shall be in accordance with Clause 31F of the main contract and tenders on a fixed price basis shall be in accordance with Clause 31B of the main contract.

See Figure 113 as an example.

These clauses will not concern us at this stage but another important point is the time allocation of the project since this will dictate the size of the labour force necessary to complete the work in that time.

Assuming that all the standard procedures for checking the suitability of an enquiry have been implemented including discussion with management in respect of type of works, period of tender, competition, proposed programme of works and also the submission of documents to one's Legal Department for appraisal, the next important requirement is to establish the price for the whole job.

The sequence of estimating is as follows:

1   Contract conditions check list. (See Appendix, page 150.)
2   Segregate project into identifiable sections.
3   Send out written enquiries to suppliers/sub-contractors for specified materials.
4   Lift off quantities.
5   Prepare cost sheets and labour running times.
6   Prepare section summary sheet.
7   Calculate 'all in' labour hourly rates.
8   Prepare preliminary sheets.
9   Complete section summary sheet.
10   Complete tender summary sheet.

TENDER SUMMARY   ELECTRICAL INSTALLATION

	*Fluctuating price	Fixed price
1 Mains and switchgear	£ 801-00	£ 881-00
2 Lighting systems	£ 5155-00	£ 5671-00
3 External lighting	£ 550-00	£ 604-00
4 Power supplies	£ 1025-00	£ 1127-00
5 Clock circuits	£ 306-00	£ 337-00
6 Ventilation installations	£ 897-00	£ 987-00
7 Fire alarm system	£ 1575-00	£ 1732-00
8 PC sums	£ 850-00	£ 850-00
9 Temporary lighting and power	£ 828-00	£ 912-00
10 Cable duct installation	£ 1560-00	£ 1716-00
	£ 13547-00	£ 14817-00

Labour percentage .... 25% ....
Material percentage .... 75% ....

* Fluctuating price base month April 1982

Figure 113   *A tender summary showing fluctuating price and fixed price quotations for the same job*

Briefly, the senior estimator checks through the tender documents/drawings and from these completes the contract conditions check list. This is then referred to the management so that a decision can be made whether to prepare a tender. If the decision is taken to tender results, then the conditions of contract are sent to the Legal Department for perusal with a request to return the documents at least seven days before the tender is due for submission.

Prior to commencement of the estimating procedures, the project drawings should be analysed and segregated into readily identifiable sections in order that the tender can be readily used for cost control, should the offer be successful. Such sections may be required to suit the client's tender summary or alternatively for choosing specific boundaries such as 'plant rooms', 'external mains', 'basement', 'ground floor' etc. Where a multiple of services are involved, for example, heating, plumbing or electrics etc., each is to be dealt with separately for various sections of the project.

Written enquiries are sent out to various suppliers/sub-contractors, inviting them to quote for the equipment/services in question. Such enquiries must be fully detailed and wherever possible the relevant extracts of the specification should be copied in order that there can be no 'misunderstanding' of the requirements. In this respect a careful check is to be made of the specification in order that any clauses of a general nature are also copied for the supplier/sub-contractor, that is, testing, painting, special delivery requirements etc. Tenders should be advised on the method of recovery of increased costs, that is, net recovery NEDO formulae (National Economic Development Office price adjustment formulae) and in respect of the latter the weightings for labour/material should be stated - together with the formulae being used, that is, 'specialist engineering installations'. On no account should the formulae for other industries be accepted (such as 'thermal insulation') and where a supplier/sub-contractor incorporates such in his offer then the offer should be rejected. Where a *fixed price* tender is required, the supplier/sub-contractor is to be notified of the fixed price period. The enquiry letter in this instance should indicate: location of site (by town or area of town, not job name); date tender to be in *our* hands. Where an offer is received which has special conditions of payment, that is, pro-forma invoice, 75 per cent before delivery etc., the estimator is to bring this to the attention of the management.

With regard to lift off quantities, these are lifted off the drawings using the appropriate scale. The quantities referred to with all necessary plant/equipment etc. should be transferred to the tender cost sheets using a logical layout for ease of checking, that is, group with items of plant/equipment sundry items as are applicable to the specific plant, for example:

1 Fuseboard (plant item)
   Fixings
   Fuses
   Phase identification
   Labels/circuit charts
2 Oil tank (plant item)
   Anti-friction pads
   Oil gauge
   Oil heater
   Sludge valve
   Alarm device (where applicable)

All materials should be priced at net cost (except for 2½ per cent monthly account discounts) and preferential discounts should be deducted where applicable. Wastage should be allowed on all pipework, conduit, electrical trunking, cable trays etc. to allow for unusable offcuts.

Labour running rates used are to be in accordance with the company standard data books and adjustments up or down made in respect of special site conditions, that is, allowance for occupied buildings, existing building, spread out site etc., or repetitive work, prefabrication. They must be agreed with management. Any additional factors are to be included in prime cost. Cranage, off loading, attendance, carriage and cartage are to be included immediately following the total material costs or total labour hours. On completion, all cost sheets are to be checked and stamped by the comptometer operator.

On the section summary sheet, enter the material costs and labour man hours (totals) for each section of the project. Check that all sections are incorporated to deal with the whole of the project. Insert in 'other adjustments' boxes, any adjustment necessary (see Appendix, page 152).

Complete calculations of labour cost per man hour using information on basic wages, fares, allowances, expenses, travel time etc. Arrive at estimated labour cost total and check stamp by comptometer operator.

Prepare preliminaries sheet for costs involved in setting up, running and clearing site. Calculate labour costs using total hourly rate as described above. Check and stamp by comptometer operator (see Appendix, page 154).

Complete section summary sheet as follows:

(a) Insert preliminaries total in the box at the bottom of column 1. Sub-divide this sum under various sections in column 2 (excepting provisional or PC sum items). Sub-division of these preliminaries may in certain instances require non-proportionate allocations to the various sections.
(b) Enter total material costs by adding columns 1, 2 and 3 (where applicable).
(c) Using labour rate, calculated as above, complete the box at the bottom of column 5, headed 'Labour Costs' (excluding preliminaries).
(d) Spread labour preliminaries and enter against sectional items in column 6 (excepting provisional or PC sum items). Similar to material preliminaries, the allocation of labour preliminaries may not necessarily be proportionate to the sectional labour costs.
(e) Complete total labour cost column by adding columns 5, 6 and 7 (where applicable).
(f) Complete total prime cost column by adding columns 4 and 8.
(g) Agree with management gross profit to be added to prime cost. Insert percentage at top of column 11.
(h) Complete remainder of the section summary sheet.

Check and stamp by comptometer operator.

The next step is to transfer total material cost and total labour cost from the section summary sheet to the tender summary sheet and complete this part. The total tender figure should match the total tender figure as shown on the section summary sheet. Again, check and stamp by comptometer operator. The manager/director must then sign tender summary sheet on completion.

Where it is necessary to produce priced bills of quantities, agreement should be reached with management as to the format to be adopted. In some instances it may be to advantage to produce priced bills of quantities to segregate preliminaries and show this as a lump sum, as this should improve cash flow and also provide a basis for a prolongation claim.

Where it is necessary to produce a schedule of rates, agreement is to be reached with management on the basis to be used. In most instances such rates will become the basis for variations and in consequence these will not be based on the gross value of the estimated costs due to the following factors:

(a) Smaller quantities will generally be more expensive per unit than the larger quantities that are in the main estimate.
(b) There is a higher overhead cost in buying, invoicing etc. for smaller quantities. In this respect it costs approximately the same overhead to purchase 4 m of 25 mm tubing as it does 40 000 m of the same size tubing.
(c) Delivery charges are often made for smaller deliveries whereas bulk deliveries are made free of cost.
(d) The handling costs on site are higher per unit for small deliveries.
(e) Variations often involve additional overheads in respect of ascertaining actual requirements, supervisory costs, preparation of estimates etc.

In estimating the labour costs for the electrical work, the majority of contracts are worked away from the firm's main office and because of this, due consideration should be given in assessing fares and allowances/travelling time based on normal agreements. Where necessary, allowances must be made in respect of height money, working in exposed conditions, dirty money or the special rates applicable for large engineering construction sites. These are to be calculated by adding a suitable percentage to the basic wages per man, per week. All non-productive time for chargehands, foremen, storekeepers etc. should be included in preliminaries.

From this procedure a typical quotation can be

made. With regard to conditions of tender, the following are typical clauses likely to be met:

Builder's discount
(a) 'This tender includes a 2½ per cent discount for the Main Contractor which shall be payable if payment is made in accordance with the terms and conditions of the sub-contract.'
(b) 'This tender is strictly net and where a cash settlement discount is required for a Main Contractor, this should be added to the figure/s shown.'
(c) 'This tender is strictly net.'

Item (c) above would only apply where one is entering into a direct contract.

VAT
'This tender is exclusive of any Value Added Tax chargeable on the work. Our tender figure will be increased by such amounts as are applicable.'

Main contractor
'This tender is subject to confirmation on the appointment of the Main Contractor.'

Types of sub-contract document
(a) For use when one is nominated.
    'This tender is submitted on the understanding that the sub-contract to be entered into will be the NFBTE/FASS/CASEC "Green Form" latest edition.'
(b) For use where one is not nominated.
    'This tender is submitted on the understanding that the sub-contract to be entered into will be the NFBTE/FASS/CASEC "Blue Form" latest edition.'

Confliction of sub-contract and specification
'This tender is submitted on the understanding that where there is a confliction between the terms and conditions of the sub-contract, that the terms and conditions of the sub-contract will apply.'

Fixed price
'This tender is submitted on a fixed price basis for a period of . . . months from the date of tender and is subject to receipt of your order and full instructions within . . . days from today's date. The tender price shall, however, be subject to adjustment to meet any fluctuations in cost arising from Government Orders, Regulations or other leglislation occurring between the date of the tender and the completion of the work. Should

the contract period be extended beyond the stated completion date, i.e. . . . . (insert date) we would require to be reimbursed for any additional cost incurred in labour, materials, preliminaries, overheads and profit.'

Variable price

Where formula method of calculating increased cost applies, then:

'Our tender is submitted on the understanding that our contract price will be subject to fluctuations in accordance with the relevant NEDO formulae as Clause 23F of the NFBTE/FASS/CASEC – "Green Form" (or Clause 23F of the "Blue Form" where applicable).'

Programme

'Our tender is submitted on the understanding that subject to the receipt of information our works will be complete by . . . (insert date) and that a detailed programme for our work will be drawn up and be subject to agreement between ourselves and the Main Contractor.'

Working hours

'The tender allows for our work to be carried out during the usual recognized working hours for this industry including customary overtime, and we assume that we will be allowed full access to the site and use of site facilities when required. If overtime during evenings, nights or weekend periods is worked on your instructions, the additional cost of "added time" will be charged.'

Liquidated damages

(a) 'Our tender is submitted on the understanding that liquidated damages will not apply to our work.'

or

(b) 'Our tender is submitted on the understanding that liquidated damages will not apply as this is adequately provided for in Clause 8 (a) of the sub-contractor's document issued by the NFBTE/FASS/CASEC or Clause 10 (2) in lieu of 8 (a) where the sub-contractor is under the Blue Form.'

Supply of materials and labour

(a) 'This tender is conditional upon Clause 23 (j) (i) and (ii) being included in the standard RIBA Form of Main Contractor.'

(b) 'It is observed that Clauses (j) (i) and (ii) are deleted and in this respect where supplier/sub-contractor is named in the specification. We reserve the right to require a change of such nominated supplier/sub-contractor where they cannot meet the required programme of works. Any such change to be dealt with as a variation in accordance with Clause 11 (2).'

Insurance

Where the conditions of contract are RIBA the following applies:

'Notwithstanding anything to the contrary in the specification and/or tender documents, our tender is submitted on the understanding that the insurance requirements as far as we are concerned are as detailed in Clause 4, 5 and 6 of the standard form of sub-contract issued by NFBTE/FASS/CASEC.'

In instances of other types of contract, you will be advised by the Legal Department of what is required.

Site facilities

This tender does not include for:

(a) All builders and the associated works of other trades.

(b) The supply, erection and positioning of all scaffolding, ladders, movable platforms or hoisting and lowering gear, necessary for placing in position and erecting plant and equipment properly on site for use on the contract.

(c) The cost of operating any part of the installation, other than attendance necessary for commissioning and testing.

(d) The cost of any fuel, water or electricity used on site.

(e) The off-loading and placing in store of such materials as are delivered to site in advance of erection starting date.

(f) The off-loading, hoisting and placing in position all plant equipment and materials relevant to the works.

(g) Any fees, incidental to the work which are compulsorily payable by reason of any statute, bye-law or regulation.

(h) Protection of all materials and goods after they are fixed.

(i) The provision of supply authorities services, that is, water, gas, electricity, soils, drains etc.

Design

Delete any of the forgoing items not applicable and add anything of a special nature as may apply in a particular instance.

Either
(a) 'We have included for and based our tender upon plant duties and equipment detailed in the engineer's specification and drawings. Our responsibility covers the procurement and installation of the plant and equipment as specified. We cannot accept any responsibility for inadequacies or defects in the specification.'
or
(b) 'In the limited time available for tendering, we have not been able to check any calculations or design aspects of the installation and have quoted on the basis of materials shown and specified. If our offer proves acceptable, we would be pleased to discuss this matter further with you.'

Note: Any further developments in respect of (b) are to be referred to management in order that a policy decision can be made in respect of our willingness or otherwise to accept any design commitment.

Accoustics
'With reference to the performance requirements in respect of noise levels, we wish to advise you that our offer is based only on the materials and plant as set out in your specification. We would be pleased to discuss this matter further with you to consider whether it is necessary to make any variations to meet your requirements.'

Existing installation
'Our tender is submitted on the understanding that the existing installation and equipment which are being re-used, operate satisfactorily and are in sound condition. Should it be found that this is not so, then any remedial or additional work that is involved would be subject to a variation to the contract sum.'

Compliance with local authorities
'In the time available for tendering, we have been unable to obtain outline approval from the Local Authority and other statutory bodies to confirm whether or not the plant and equipment meet with their requirements, but we have based our proposals on our previous experience of what is normally required to obtain the necessary approvals.
In the unlikely event of modifications being necessary in this respect then these would be discussed with your goodselves together with any financial considerations.'

Maintenance of plant and temporary operation
(a) 'During the defects liability period (that is, . . . months), our responsibility will exclude any routine maintenance and attendance of the installation or any liability for damage caused by the lack of proper maintenance/attendance.'
(b) 'This tender excludes any costs involved in operating the services before final handover of the completed installation. If prior operation is required, the costs involved will be charged as an extra to the contract sum.
We have also assumed that the necessary insurance policies for the installation will be undertaken by others at no cost to us.'

PC items/provisional sums
(a) 'We have assumed in our tender that any prime cost sum includes the cost of delivery, off-loading, handling, site erection and commissioning.'
and/or
(b) 'Our tender is based on the understanding that we shall be able to place contracts with the nominated sub-contractors/suppliers under the terms and conditions of the main contract.'

Form of warranty
(a) 'We regret that we cannot accept any amendments to the RIBA warranty where these are contrary to the agreement reached by the respective parties to the agreement. We have, therefore, deleted such amendments.'
and/or
(b) 'With regard to Clause A (2) of the Employer/Sub-Contractor form of agreement, our tender is submitted on the understanding that a detailed phased programme and completion date for our work will be drawn up and be subject to agreement between ourselves and the Main Contractor.'

Dayworks
Any additional works that are not the subject of a separate quotation will be charged at the cost of these additional works on the basis set out below. The prime cost will be calculated in accordance with the definition agreed between RICS and . . . (insert HVCA/ECA etc.) and current at date of tender.

Percentage additions to prime cost
1   Labour
2   Materials
3   Plant

Where we are required to give a 2½ per cent cash settlement discount to a Main Contractor 1/39th is to be added.

The above rates are exclusive of VAT.

Variations

'This tender is based on the assumption that where variations to the works are priced on the unit rates, such rates will apply, providing always that the variations are carried out in the normal sequence of working and under the same conditions and circumstances as the sub-contract.'

Bill of quantities

'We wish to advise that irrespective of any discrepancies between the information given on the drawings, specification and bill of quantities, we have prepared our tender exactly in accordance with the quantities described in the bill of quantities.'

Limit of variations

'If the net effect of all variations (other than those arising by reason of any clause relating to variations in price of materials and/or labour) shall be found during or on completion of the works to result in an addition or a reduction greater than 15 per cent of the sum named in the tender price, the amount of the contract price shall be amended by a sum which shall take account of any additional costs caused by the changed circumstances of the work to be carried out.'

Asbestos

Applicable to existing buildings.

'In the event of asbestos products being discovered in the premises which involve us in implementing the special requirements necessary under current leglisation, the contract sum shall be amended to take account of the additional costs incurred.'

Miscellaneous points

1   A *provisional sum* is the amount which the client is prepared to spend on certain unnamed facets of the installation equipment. It will include the contractor's costs, overheads, profit and discount to the Main Contractor where applicable. Often, the words 'provisional sum' describe an amount to be spent or deducted at the discretion of the architect, and on completion of the contract the total sums included are deducted from the account and in place inserted the actual cost of works performed against such provisional sums.

2   The words *'prime cost'* or *PC sums* describe amounts set against specific items for which the architect will have previously obtained quotations from manufacturers of these items. He may then include the cost of these items in the enquiry to the builder. It is therefore important for these *known costs* to be incorporated in the Tender Summary (see Appendix) while at the same time the installer must include an amount elsewhere to cover the cost of installing the items.

3   A Main Contractor is entitled to receive a discount, that is, *Main Contractor's Discount* whenever he pays his sub-contractor. This is usually added to the price to be quoted (less provisional and contingency sums). This is the accepted 2½ per cent as 1/39th (see Appendix).

4   A *bill of quantities* is a list of all the components of an installation which have been compiled by a professional quantity surveyor. Here a tenderer is required to price individually on a supply and install basis. In practice, where the contractor has submitted a 'rate price', that is, a price for supplying and installing one unit of a particular item, the total tender price is found by adding the required number of each of the unit prices.

5   *Schedule of rates* are unit costs appearing in a bill of quantities. They are normally required for abstract items which may be appropriate or not inappropriate to the works but none the less are produced to give a total cost. A schedule can be used for evaluating variations to the contract.

6   Where a tenderer is required to submit a *fluctuations* price, the fluctuations are calculated in accordance with the National Economic Development Office (NEDO) Price Adjustment Formulae for specialist Engineering Services. The formulae uses standard indices for labour and materials to which weightings are applied. These indices are calculated monthly and published in *Monthly Bulletin Construction Indices* – published by The Property Services Agency (PSA).

The indices for labour for use in connection with the formulae reflect the price level obtaining on the first day of the calendar month to which they refer. The indices take into account authorized variations in the rates of wages etc. as agreed and published in the various National Agreements and Rules.

Revision exercise 5

1   Explain the following terms:
    (a) project co-ordination
    (b) network analysis
    (c) bar chart
    (d) specification
    (e) fixed price contract

2   A bar chart for a small project is shown in Figure
    114. It must be noted that activities D and E can
    only commence when B is complete, activity F
    can only commence when A is complete, acti-
    vities G and H can only commence when D is
    complete, activity I can only commence when E,
    F and G are complete and activity C does not
    restrict other activities.
    (a) Prepare a network diagram of the
         programme.
    (b) Clearly mark on the diagram the critical
         path.
    (c) State how much longer would be spent on
         activity E without increasing the total
         project time.

                                              CGLI/C/82

3   (a) What is meant by a site diary?
    (b) What is the use of such a diary?
    (c) Write *four* typical entries that would be
         found in a site diary.

                                              CGLI/C/81

4   (a) Explain clearly the difference between an
         electrical contract being carried out as a
         'lump sum contract' compared to a contract
         carried out on a 'bill of quantities'.
    (b) Describe the administrative actions that
         should be taken by the electrician in charge
         if the contract is being conducted on a bill
         of quantities.

                                              CGLI/C/82

5   Explain the difference between 'estimating' and
    'tendering'.

6   State *three* important factors likely to be con-
    sidered by management when deciding whether
    to submit a tender for a particular project.

7   Prepare notes for guidance to be given by a site
    foreman to a job/shop representative (shop
    steward) in order to ensure good labour relations
    on a site. The areas to be covered should include:
    (a) clocking on and off
    (b) tea and meal breaks
    (c) washing and changing facilities
    (d) grievance and dispute procedures
    (e) meetings of men
    (f) redundancy procedure

                                              CGLI/C/77

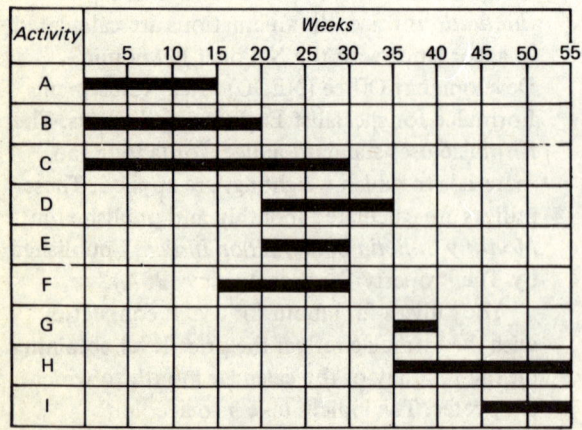

Figure 114

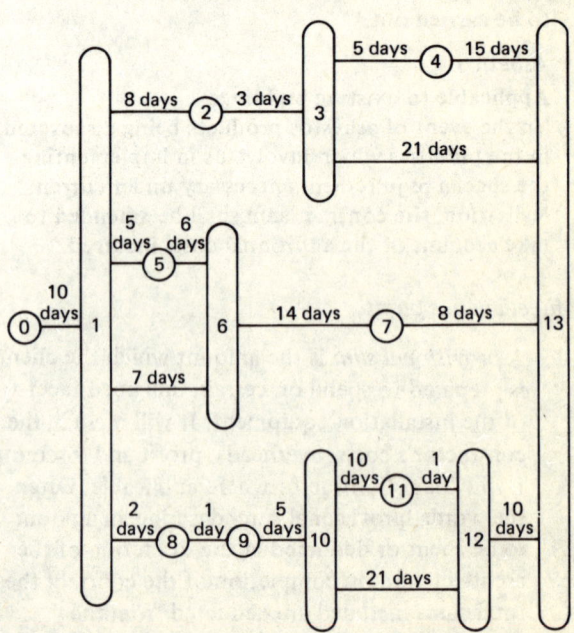

Figure 115

8   Figure 115 shows the network analysis of a contract about to be undertaken by an electrical firm. The length of time required to complete each activity is shown in days.
(a) Determine the activities forming the critical path and draw the bar chart for this network.
(b) State (i)   the activities on the critical path
          (ii) the length of time needed to complete the contract
          (iii) the effect of activity 6 to 7 taking seven days longer than expected.
                                                    CGLI/C/77(Mod.)

9   An electrician in charge of the electrical installation work on a new project has to organize a section of work which can be broken down as follows:

Installing circuit conduits	40 person/days work
Installing submain cables	12 person/days work
Installing distribution gear	4 person/days work
Drawing in circuit cables	8 person/days work
Fixing/connecting accessories	4 person/days work
Fixing/connecting luminaires	8 person/days work

The following constraints apply:
    Not more than four workers can be employed at any one time.
    Work on submains must start when 50% of circuit conduit is installed.
    Other work can proceed in a logical sequence when 50% of the required previous work has been completed.
Draw a bar chart of the programme for the above work so that it may be completed within the minimum possible time. Indicate on the chart the number of workers employed at each time.
                                                    CGLI/C/84

10 The electrical installation in a new small office development comprises fluorescent lighting, general-purpose socket outlets, small power associated with the heating system and a fire alarm system. The wiring will be PVC cables drawn into screwed metal conduits, the latter laid in the floor screed and buried in plastered walls. It may be assumed that:
    (i)   the electrical work is scheduled to be completed in exactly six months
    (ii) work can proceed uninterruptedly during the six month period
    (iii) six deliveries of material will be made at monthly intervals
(a) Explain why it is not advisable to arrange for all the materials needed for the job to be delivered together at the commencement of the work.
(b) Set out an appropriate delivery schedule for material indicating to the purchasing department of the contractor's organization when particular material will be required.
    The schedule may be set out in the form of:
    Delivery No. 1, X% conduit, Y% conduit accessories.
                                                    CGLI/C/78

chapter six

Electronics

After reading this chapter you will be able to:

1 Describe the operation of a cathode ray oscilloscope and carry out calculations to determine trace magnitude and frequency.

2 Explain with or without diagrams the operational functions of a number of listed transducer devices.

3 Explain with or without diagrams the operational functions of a number of listed semi-conductor devices.

4 State applications for a number of listed transducer and semi-conductor devices.

5 Distinguish between AND, OR, NAND, NOR and NOT logic gates and draw diagrams of truth tables associated with logic gates.

Cathode ray oscilloscope

This is a piece of electronic test gear mainly used to look at waveforms and voltage levels of circuits. Essentially it comprises a conical-shaped evacuated glass tube in which a beam of electrons is guided towards a fluorescent screen. When the electrons hit the screen, the screen glows and the variable being displayed can be measured in the form of a graph to a base of time. The construction of the tube and its internal components is shown in Figure 116. It can be seen that the three main components are:

(a) the electron gun;
(b) the deflection system;
(c) the fluorescent screen.

The purpose of the electron gun is to produce the stream of high-speed electrons. This part consists of a *heater*, a *cathode* and a *control grid*. The source of electrons is provided by the heated cathode which takes the shape of a nickel cylinder which is oxide coated and closed at one end. Surrounding the cathode is another much larger cylinder, the control grid. This is opened at the cathode end and closed at the other end except for having a very small aperture. This constriction serves to concentrate the electrons into a beam before they pass through the first accelerating anode (A1). This anode and anode A3

are in the form of discs each having a small central hole to enable the electrons to pass. Between these two anodes is a second anode (A2) in the shape of a cylinder and its function is to focus the electrons into a narrow beam to show up as a fine spot on the screen.

In order for the spot to trace any waveform on the screen, it must be movable. This is the function of the deflection system, that is, the X and Y plates. They have the job of moving the beam in two directions. The first set of plates, that is, the Y plates, deflect the beam vertically and the second set of plates, that is, the X plates, deflect the beam horizontally. Figure 117 shows the effects of *electrostatic deflection*. It should be pointed out that if no potential is applied between the plates, then the electron beam will strike the centre of the screen. Thus, movement of the beam is made by varying the voltage applied to the deflection plates. Another method of moving the beam is by magnetic field.

With regard to the screen, its purpose may be two-fold, the fluorescent coating converts the electron beam into light and it may also allow the trace to remain after the initial bombardment. This latter point is important when one is examining traces, particularly in radar work, where the image is required to remain until the next scan. It will be noticed in Figure 116 that the screen is connected to anode A3

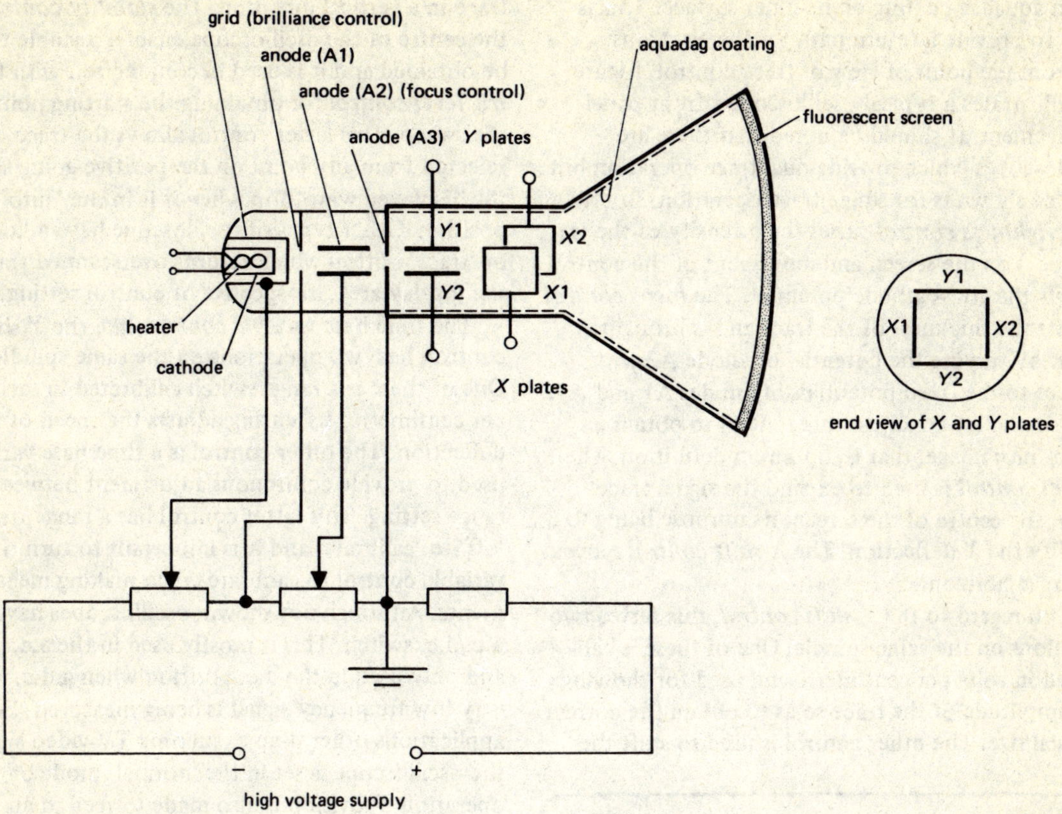

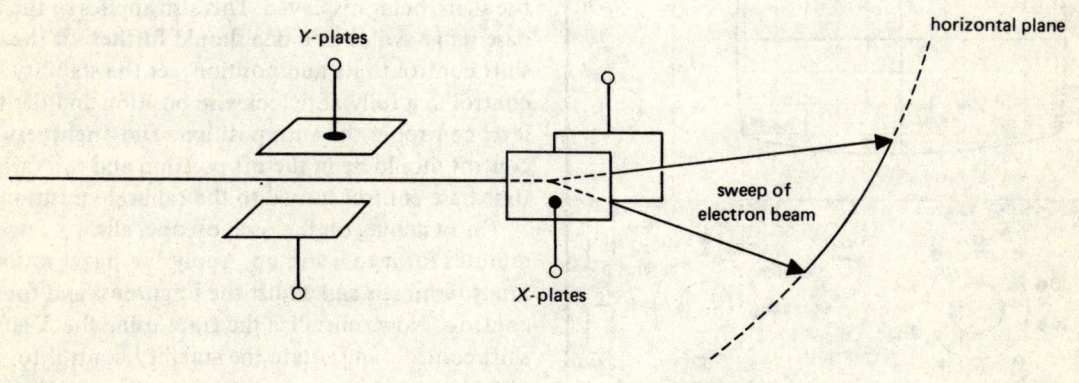

grid (brilliance control)
anode (A1)
anode (A2) (focus control)
anode (A3)
Y plates
aquadag coating
fluorescent screen
Y1
Y2
X2
X1
heater
cathode
X plates
Y1
X1
X2
Y2
end view of X and Y plates
–
+
high voltage supply

Figure 116 *Cathode ray tube*

Y-plates
horizontal plane
sweep of
electron beam
X-plates

Figure 117 *Electrostatic deflection by varying potential on X-plates*

via an aquadag coating on its inner surface. This is done to provide a return path for the electrons.

From the point of view of trace control, Figure 118 illustrates a typical oscilloscope's front panel arrangement. It should be noted that there are oscilloscopes which provide dual trace operation but the one shown is for single trace operation. Briefly, the *brightness control* varies the intensity of the trace (or spot) on the screen and movement of the control adjusts the grid–cathode potential. The *focus control* adjusts the thickness of the trace and is brought about by varying the potential of anode A2 with respect to the fixed potentials of anodes A1 and A3. One should try to adjust this control to obtain a sharp trace image, that is, optimum definition. The *X gain control* is used to expand the signal trace about the centre of the screen, its purpose being to amplify the X deflection. The *X shift control* moves the trace horizontally.

With regard to the *Y shift control*, this serves two functions on the same spindle. One of these is calibrated in volts per centimetre and used for showing the amplitude of the trace so as to obtain the correct vertical size. The other control is used to shift the

trace in a vertical direction. The *stability control* in the centre of the oscilloscope enables a stable trace to be obtained and it is used in conjunction with the *trig level control* for obtaining the starting point of the sweep. This latter control allows the trace to be selected from any point on the positive going slope of the displayed waveform when it is in the 'auto' position. Other types of oscilloscope have a 'locate' or 'trace' button which returns overscanned traces to the display area, irrespective of control setting.

The time base *variable control,* like the Y shift control, has two operations on the same spindle. One of these is a range switch calibrated in seconds per centimetre. Its setting adjusts the speed of the X deflection. The other control is a time base variable used to provide continuous adjustment between each range setting. This latter control has a range from 'off' to 'calibrate' and it is important to turn the variable control to calibrate when making measurements. Although not shown, oscilloscopes have an a.c./d.c. switch. This is mostly used in the a.c. position and only used in the d.c. position when a d.c. level or very low frequency signal is being measured. For all applications other than examining TV video signals, the oscilloscope is set in the 'normal' mode of operation. Provision is also made to feed in an external trigger signal.

With regard to operation, before switching on, one should set the X gain control to its minimum position, the X shift to its mid position and set the volts/ centimetre control to a suitable value in relation to the signal being displayed. This also applies to the time base range switch and one should further set the Y shift control to its mid position, set the stability control in a fully anticlockwise position and the trig level control in the auto position. The brightness control should be in the off position and the variable time base control turned to the calibrate position.

On switching on the oscilloscope, allow a few minutes for it to warm up. Apply the signal to the input terminals and adjust the brightness and focus controls. Now centralize the trace using the X and Y shift controls and rotate the stability control to obtain a stable image on the screen. The amplitude of the trace is found by direct measurement of its peak-to-peak value in centimetres and multiplying this by the setting of the volt/cm control switch. So, for example, if the peak-to-peak signal was 4 cm on the

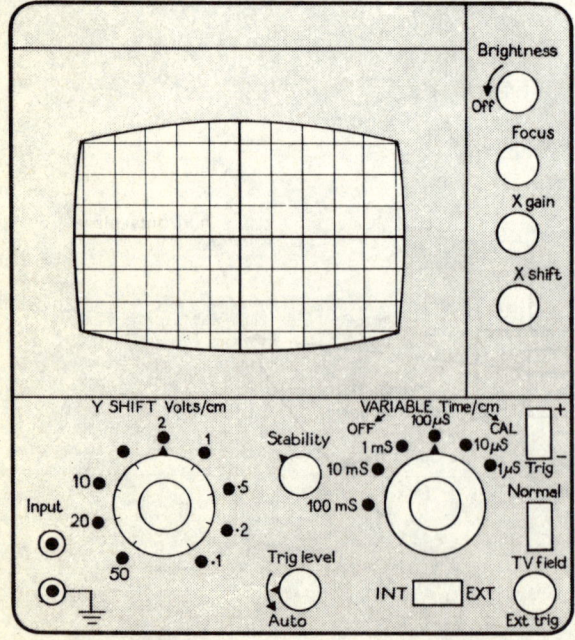

Figure 118   *Front panel of a typical oscilloscope*

oscilloscope's graticule scale and the volt/cm control was set at 20 volts/cm, then the amplitude of the trace would be 80 V.

Where frequency is required, this is calculated by using the time base control in a suitable position and measuring the horizontal distance it takes to complete cyclic variations. Frequency is then found by

$$F = \frac{\text{number of cycles}}{\text{time/cm x number of cm}} \text{ Hz}$$

If 2 cycles were spread over 6 centimetres and the time base control set to 10 milliseconds, then the frequency will be 33.3 Hz.

Transducers

These are devices used for converting one form of energy signal into another, and although this description could easily fit a wide range of mechanisms, even the cathode ray oscilloscope previously mentioned, it

is meant to refer to devices of specialized application, particularly those in the electronic industry, field of instrumentation and control systems.

Figure 119 shows the principles of a two-stage transducer. The first section of the device is called the mechanical transduction stage which translates the *measurand* (that is, the quantity, condition or property which is measured) into a displacement or stress which acts as a stimulus for the second section. This second section is called the electrical transduction stage where the required output signal is produced. The following types of transducer will be briefly discussed:

(a) photovoltaic cell
(b) piezoelectric transducer
(c) thermoelectric transducer
(d) variable capacitor transducer
(e) variable resistance transducer
(f) light dependent resistors
(g) photodiodes
(h) phototransistors
(i) strain guages
(j) thermistors

The *photovoltaic cell* is one of several photo-detector transducers which respond directly to incident photons (that is, uncharged particles). The measurand is converted into a change in the voltage generated when a junction between certain dissimilar materials is illuminated. In other words, the photo-voltaic cell generates a potential difference when it is

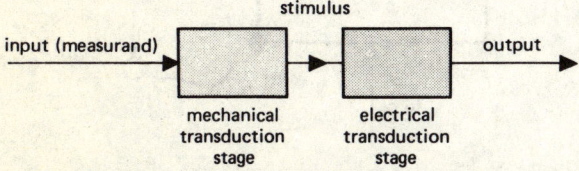

Figure 119   *Two-stage transducer*

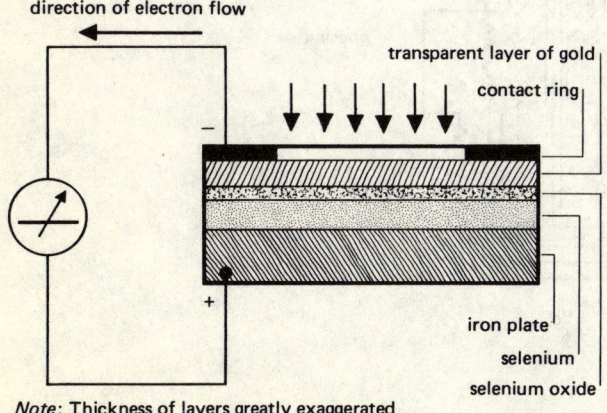

Note: Thickness of layers greatly exaggerated

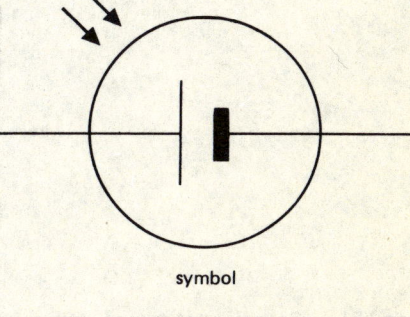

Figure 120   *Photovoltaic cell*

exposed to light. The cell consists of an iron plate coated with a thin layer of selenium. The upper surface of the selenium is oxidized and covered with a gold film, so fine that it becomes transparent to light. In contact with the gold film is a metal connecting ring (see Figure 120). When light falls on the surface of the cell, it passes through the transparent layer of gold and causes electrons to be released from the surface of the selenium. Because selenium is a semiconductor material, electrons flow through the oxide layer to the gold and the gold acquires a surplus of electrons (that is, it produces a negative charge). The selenium and iron plate acquire a positive charge. Application of the voltaic cell is in portable lightmeters, solar panels and moving coil instruments.

In the *piezoelectric transducer,* the measurand is converted into a change in the electrostatic charge and voltage generated by certain crystals when mechanically stressed through either bending, compression or tension force. Ceramic piezoelectric material is quite often used today as the conversion medium since it has good mechanical properties. Figure 121 shows a typical piezoelectric pressure transducer. Its design uses a flat diaphragm whose inside surface is directly in contact with the crystal or ceramic element. The motion of the diaphragm can cause tension/compression or bending in the element depending on design. Piezoelectric transducers have a wide application in gas ignition devices, cigarette lighters, microphones, pick-up cartridges etc.

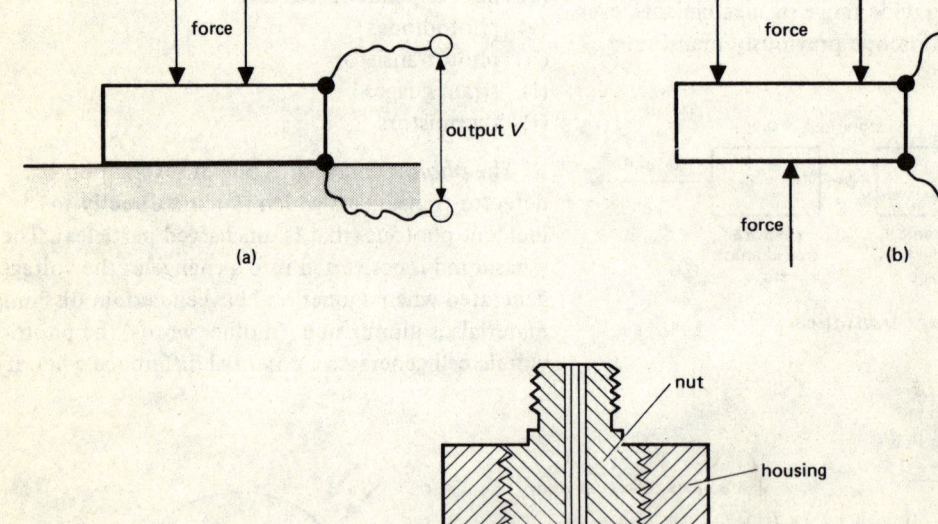

Figure 121   *Piezoelectric transduction*
                 (a) Compression/tension
                 (b) Bending
                 (c) Sectional view of piezoelectric
                      pressure transducer

A thermoelectric transducer such as a *thermocouple* is used for temperature measurement. It consists of a pair of wires of different conductor material which are joined together at one end. The materials used often depend on the range of temperature that is required to be measured. For example, iron and constantan would be used for temperatures between $-200°C$ and $+900°C$ and platinum and platinum rhodium would be used for temperatures between $0°C$ and $1550°C$. Figure 122 shows a basic thermocouple circuit. When a temperature difference exists between the sensing (hot) junction and the reference (cold) junction, an e.m.f. is produced which causes current to flow in the circuit. The cold junction temperature is measured and used to compute the value of the temperature at the hot junction. A typical thermocouple probe used for measuring oven temperatures or liquid temperatures is shown in Figure 123, its range is between $-200°C$ and $+250°C$ with an output voltage of $40\ \mu V/°C$.

In a *variable capacitor transducer* the measurand is converted into a change of capacitance simply by altering the capacitor's dimensions, that is, area of its plates, distance between its plates or even the position of the dielectric. The first and last of these methods give a change in capacitance which varies linearly with the displacement whereas a change in distance between the plates gives a capacitance which is inversely proportional to the displacement. Figure 124 illustrates these methods. They find widespread use in feedback loop, operational amplifiers as well as capacitive pressure devices used for gas and liquid measurement.

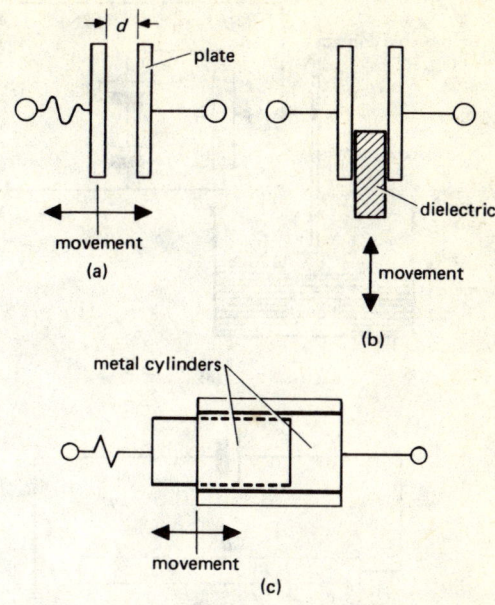

Figure 124   *Variable capacitance transduction*
(a) Movable plate changing distance
(b) Movable dielectric
(c) Movable cylinder changing area of
      overlap

In the *variable resistance transducer*, the measurand is converted into a change of resistance which can be brought about by various methods, such as, for example, by applying heat to the conductor, applying some form of mechanical strain or even by arranging the conductor so as to produce a variable resistance through a sliding contact (for example, rheostat). Figures 125(a) and (b) show how this is achieved. Both methods shown could be thought of as potentiometric transducers since the measurand is converted into a change in position of a movable contact. In Figure 125(a) it is through a change in liquid level. Figure 125(b) could represent the movable contact of a wiper arm on a resistance element.

In the field of optoelectronic devices, one comes across a number of photodetector transducers, such as *photoconductors, photo-emmisives* and even the *photovoltaic cell* already mentioned. These devices respond to incident photons (that is, uncharged particles having energy as a result of their frequency and wavelength). In the photoconducting devices the measurand is converted into a change in resistance of

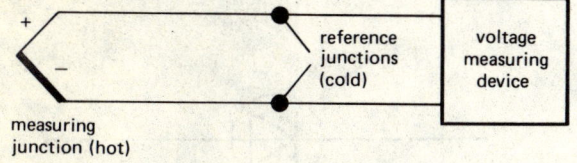

Figure 122   *Basic thermocouple circuit*

Figure 123   *Thermocouple probe*

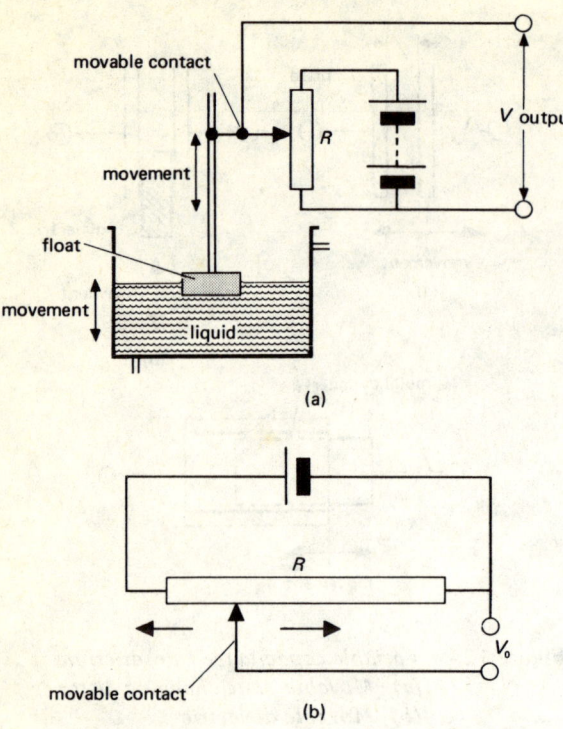

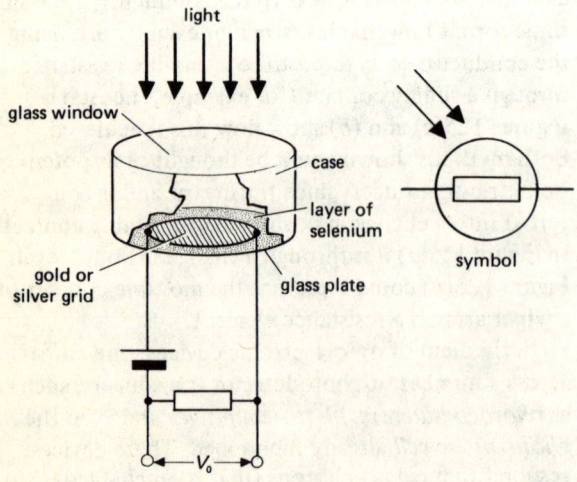

Figure 125   *Variable resistance transduction*
              (a) Simple float system
              (b) Simple wiper system

semiconductor material by a change in the material's exposure to light. A photoconductive cell is shown in Figure 126. It consists of a glass plate upon which is deposited a thin film of gold or silver in the form of two meshed grids. The grids are separated from each other by a very thin path which is covered with a film of selenium. Selenium is chosen because it has the property of decreasing in electrical resistance when it is exposed to light. When a constant voltage is applied to the cell, the current will be much greater when the cell is illuminated than when it is dark. The device finds considerable use as an automatic switch, particularly in the switching *on* and *off* of car park lighting.

The photodiode is a junction diode which utilizes the photoelectric properties. If a semiconductor diode is reverse biased, only a small current will flow. It has been found that this small current will increase if the diode junction is exposed to light. A simple circuit is shown in Figure 127. In practice, the output power from the photodiode is restricted and methods of amplification are employed to increase the power. A typical arrangement used today is shown in Figure 128. It is suitable for photometers and high speed counting.

The phototransistor is a photodiode with the addition of a collector. It is arranged so that light can shine on the base/collector junction. The current flow due to the effect of light is amplified by normal transistor action. The ability of this device to detect light and provide amplification makes it suitable for a number of 'on–off' situations such as automatic switching of light, punchcards, pattern matching and

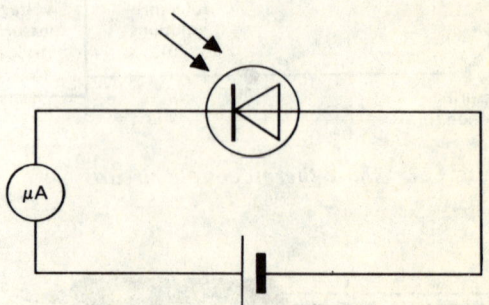

Figure 126   *Simple photoconductive device*

Figure 127   *Basic photodiode circuit*

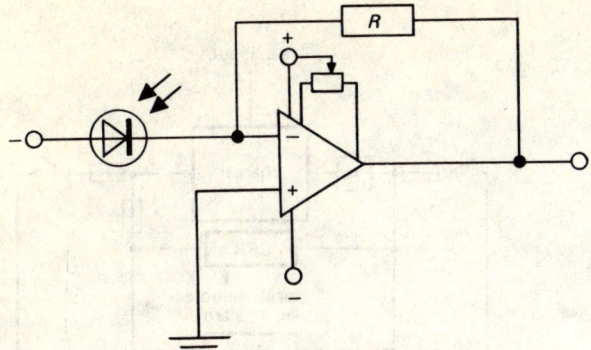

Figure 128   *Photodiode connection in a linear photometer circuit using an FET operational amplifier*

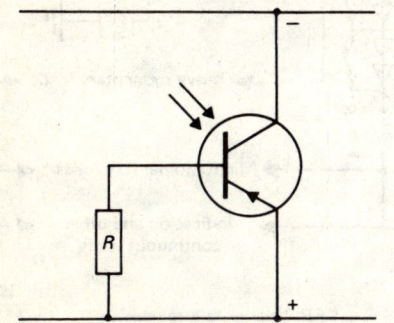

Figure 129   *Simple phototransistor connection*

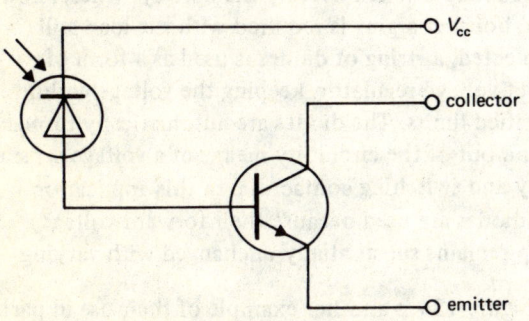

Figure 130   *Connections of a transistor preamplifier*

counters etc. Of interest, it should be pointed out that if an open circuit occurs in the base circuit of a phototransistor, the variations in collector current as a result of temperature changes could be considerable. In situations where this is experienced it is preferable to have a much higher ratio of light current to dark current (that is, current which flows in the photodiode in darkness) and this is achieved by connecting a resistor between base and emitter. Figure 129 shows a phototransistor connection while Figure 130 shows a photodiode used with a single transistor preamplifier, effectively achieving a high-grade phototransistor.

Another device worth mentioning at this stage is the *light-emitting diode* (l.e.d.). Basically, when the diode is forward biased, electrons pass into the p-region while holes occur in the n-region (see Volume 2). Some of the charge carriers, however, remain in the junction area and it is their energy which produces light. Various colours of light can be emitted using materials such as gallium phosphide and gallium arsenide phosphide. L.e.ds have a wide application.

In terms of the strain gauge transducer, which basically is a special version of the resistive transducer, the measurand is converted into a resistance change as a result of strain. The device is often used in conjunction with a strain gauge measurement bridge energized by either an a.c. or d.c. supply. The device is often stuck onto the object being measured where it converts the change in length at the surface of the material into a change in electrical resistance. Several types are in common use, namely, foil gauges, wire gauges and semiconductor gauges. Figure 131 is a sketch of a typical metal-foil strain gauge.

The last device to be mentioned is the *thermistor*. This, basically, is a temperature-sensitive semiconductor resistor usually designed from metal oxides

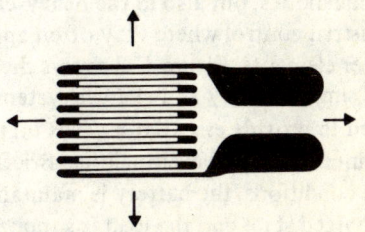

Figure 131   *Metal-foil strain gauge*

Figure 132 *Thermistor probe*

such as nickel, manganese and cobalt. The device is
used for temperature measurement and other control
applications such as trip circuits, warning circuits etc.
While the thermistor is normally a negative temperature coefficient of resistance device, a range of
positive temperature devices are on the market today.
In comparison with the resistance thermometer and
thermocouple (previously mentioned), a thermistor
has the advantages of a much higher temperature coefficient of resistance, greater sensitivity to temperature changes and a higher resistivity value. Figure
132 is a sketch of a standard thermistor probe which
has a stainless-steel body for fitting into a tank or
oven. Its temperature range is between $-10°$C and
$+200°$C.

Semiconductor devices

A description of semiconductor devices has already
been given in Volume 2 of this series. Here, it is only
necessary to mention them in support of relevant
topic areas pertaining to Course C studies. The
devices which will receive attention are:

(a) diodes
(b) transistors
(c) thyristors
(d) diacs and triacs.

From a practical point of view, semiconductor diodes
find widespread use, not only in the light-current
field of radio and television and electronics, where
they are likely to function as signal diodes and zener/
avalanche diodes, but also in the heavy-current field
of industrial control where they often appear as
rectifier elements. Figure 133 shows their use in a
typical simple battery and charger system which is
designed to provide essential services that have to be
maintained against all eventualities. Briefly, under
normal conditions, the battery is maintained in a
fully charged state and the load is supplied by the
charger. In the event of loss of supply, the con-

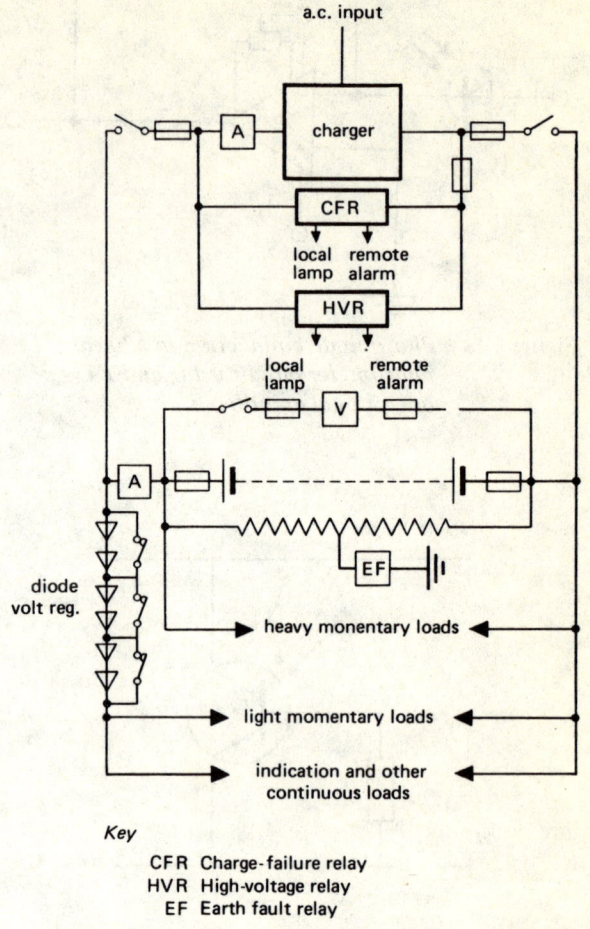

Key

CFR Charge-failure relay
HVR High-voltage relay
EF Earth fault relay

Figure 133 *Simple standby battery charger*

nected load is taken over by the battery. Where, however, boost charging is required with the load still
connected, a string of diodes is used as a form of
direct-voltage regulator, keeping the voltage within
specified limits. The diodes are automatically brought
in and out of the circuit by means of a voltage sensing
relay and switching contactors. In this application
the diodes are used because their forward voltage
drop remains substantially unchanged with varying
current.

Figure 134 is another example of their use in part
of an inverter system supplying a three-phase induction motor. The autotransformer reduces the supply
voltage which is then fed to the diodes connected as a

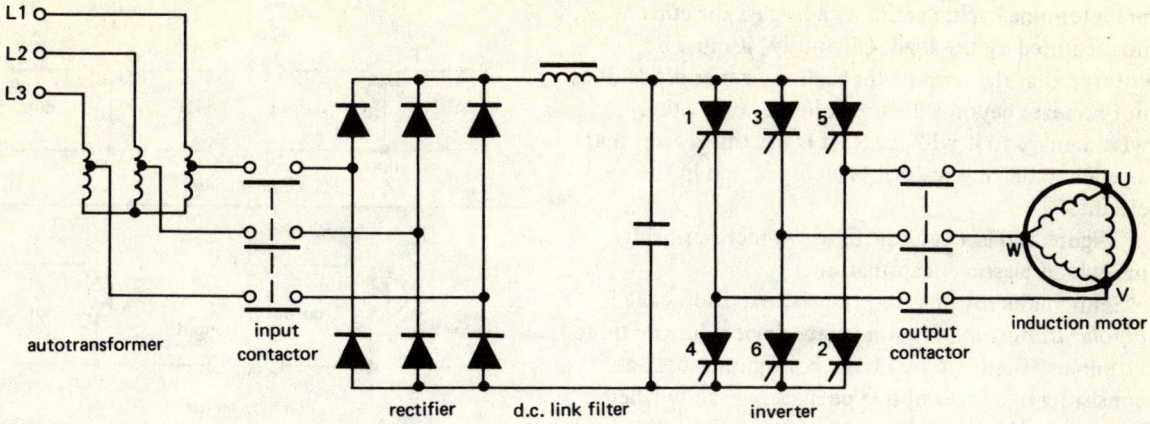

Figure 134 *Basic inverter system controlling speed of a three-phase induction motor*

rectifier via an input contactor. The d.c. output voltage from the rectifier is then fed to the inverter via an *LC* filter. The inverter uses the d.c. voltage to generate the a.c. supply for the motor via the output contactor. Figure 135 is a simple power supply unit incorporating a zener diode used for voltage stabilization. The diodes D1, D2, D3 and D4 are connected as a full-wave bridge rectifier. When one side of the secondary winding is at a positive potential, the diode connected to it (D1) will conduct and pass current through the circuit, returning the pulse through D4 and back to the secondary winding. When the supply is reversed in polarity, D2 conducts and D3 returns the pulse. In order to provide a direct current pulse at a steady value the fluctuations of the rectified current must be smoothed out. A normal smoothing circuit comprises a reservoir capacitor, smoothing capacitor and smoothing choke (or sometimes smoothing resistor). The reservoir capacitor acquires a charge in the form of pulses of current and it dissipates this charge through the load at a steady rate. When, however, the charge is withdrawn, the voltage decreases and a state of balance is achieved. This occurs when the charge is just equal to that withdrawn by the load. The effect of the smoothing resistor is to level out the rate at which current is drawn from the reservoir capacitor. Working in conjunction with the smoothing resistor is the smoothing capacitor which forms a filter circuit giving an output voltage substantially free from any ripple effects. With regard to the zener diode, this is a junction diode which can operate in the reverse breakdown region with a relatively high reverse current without damage. It is often referred to as a voltage reference diode and is non-conducting up to a

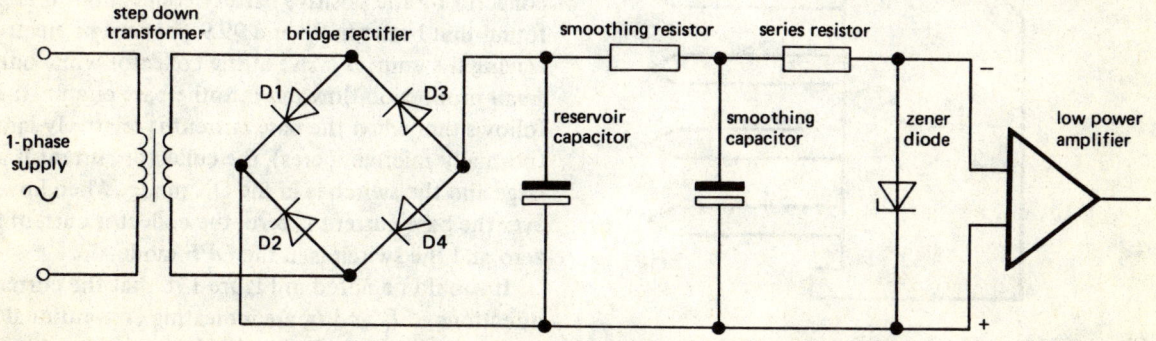

Figure 135 *Simple power supply unit supplying low power transistor amplifier circuit*

predetermined value acting as a bypass for current not required by the load. Obviously, it must be ensured that the current through the zener diode does not increase beyond the breakdown point otherwise damage to it will occur. It is for this reason that a series stabilizing resistor is incorporated in the circuit.

Figure 136 is a typical full-wave rectifier bridge module in plastic encapsulation.

Semiconductor junction transistors, often called *bipolar transistors*, are solid-state devices having three terminals. There are two basic configurations, one consisting of a layer of n-type material sandwiched between two layers of p-type material, the other consisting of a layer of p-type material sandwiched between two layers of n-type material (see Volume 2 for an understanding of p- and n-type materials). The former arrangement is referred to as a *pnp bipolar transistor* while the latter is an *npn bipolar transistor*. In practice, these devices are utilized in circuits in three different modes of connection, namely, *common base, common collector* and *common emitter*. These are shown in Figure 137.

In the common base mode, the input resistance is relatively low compared with the output resistance (say 50 Ω as against 1 MΩ). In the common collector mode, the values of input and output resistance given are reversed while in the common emitter mode the resistance values are somewhere between these two extremes (say 500 Ω as against 20 kΩ). The common emitter circuit is the most popular circuit arrange-

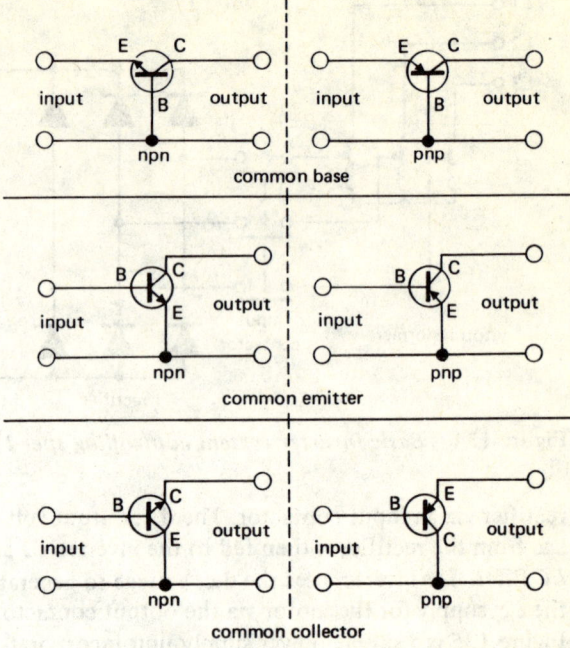

Figure 137   *Methods of connecting bipolar transistors*

ment, particularly in amplifier circuits which receive a mention later in the chapter.

For a simplified understanding of the common-emitter connections in its use as a *logic* switch, reference should be made to Figure 138. When the emitter junction is forward biased (as shown), a high concentration of electrons build up in the base region. Since the value of collector voltage is normally greater than the base voltage, the large build up of electrons in the base region is attracted to the collector by the positive battery connection. It is found that between 98 and 99.9 per cent of electrons leaving the emitter arrive at the collector while only a small proportion flow out into the base circuit. It follows that when the base current is relatively large (normally microamperes), the collector current is also large and the switch is in the ON mode. When however the base current is zero, the collector current is zero and the switch is in the OFF mode.

It should be noted in Figure 138 that the current directions of I_c and I_b are indicating conventional current flow but the action has been explained in terms of electron flow.

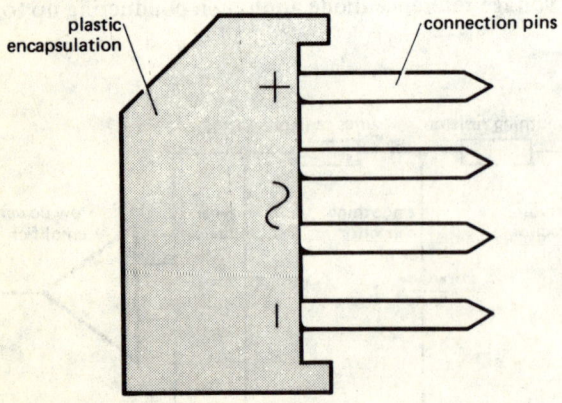

Figure 136   *Bridge rectifier component*

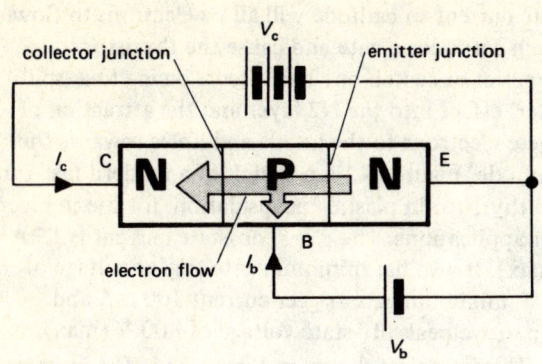

Figure 138   *Connections of a common emitter transistor*

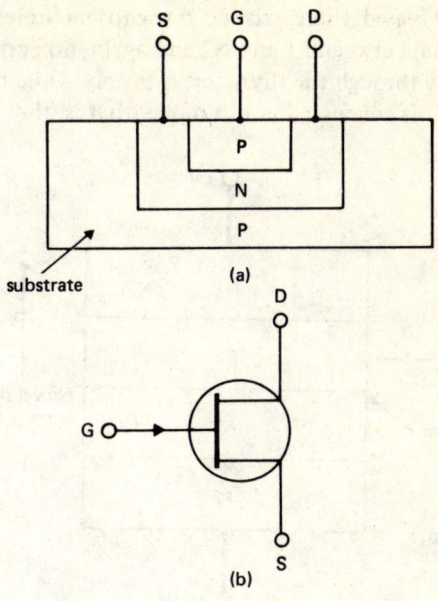

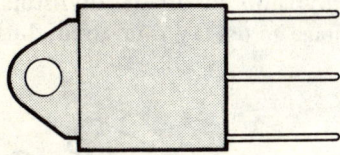

Figure 139   *n-p-n power transistor*

Figure 140   *n-channel junction FET*
       (a) Structure
       (b) Symbol

Figure 139 is a typical npn power transistor in encapsulated form for use in inverter and converter circuits, switching regulators and motor control systems.

Another transistor device requiring a brief mention is the *field-effect transistor* (FET). Unlike the bipolar junction transistor, this is a voltage-controlled device having a high input resistance. Basically it consists of either an n-type channel embedded in a p-type substrate or a p-type channel embedded in an n-type substrate. Figure 140 is a diagram of an n-channel FET. It will be noticed that it is a three terminal device. Those connecting the n-type channel are called source (S) and drain (D) while the one connected to the p-type layer is called the gate (G). It is this gate connection which is used for controlling the channel current since the gate–source junction will be

reverse biased. A number of FET variations exist but will not be discussed at this stage.

The *thyristor* or silicon controlled rectifier (SCR) is another semiconductor device and it acts as an open switch in a circuit until a switching current is applied to its gate. Once it commences to conduct, the switching current can be removed (or fall to zero) and it will continue to conduct until the load current falls to a very low level. The thyristor has many applications as an electronic switch and its gate current need only be a small value compared with its load current. Figure 141 is a diagram of its arrangement as npnp structure. When the cathode is made positive with respect to the anode, excess electrons in the N1 layer are attracted to the cathode while *holes* in the P2 layer will be attracted to the anode. This application will remove both electrons and holes from the middle P1 and N2 region of the device and consequently prevent current flowing through the thyristor. If now the cathode is made negative with respect to the anode, the holes of P1 will be attracted to the cathode and the electrons of N2 attracted to the anode. While the inner region of P1 and N2 is actually forward biased, there exist no *free* carriers (holes and electrons) between P1 and N2 and again, no current will flow through the thyristor. It is only while the thyristor is connected in this mode, that is, the

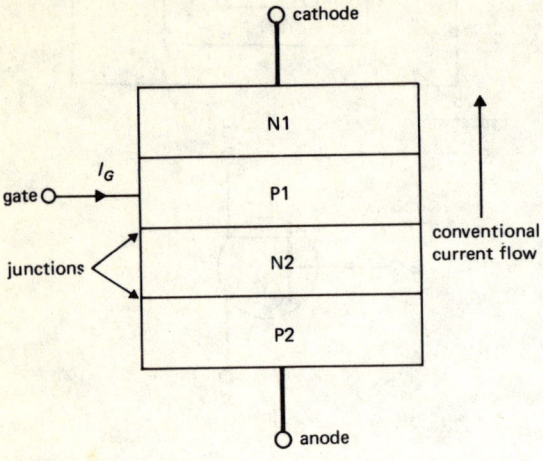

Figure 141   *Structure of an npnp thyristor*

forward blocking condition, that the application of a gate current to cathode will allow electrons to flow from cathode to gate and cause the thyristor to conduct or switch on. This occurs owing to a 'spill-over' effect into the N2 layer and the attraction of these electrons to the anode and holes towards the cathode. Figure 142 is a sketch of a modern fast-turn-on thyristor in plastic encapsulation, for use in inverter applications. The r.m.s. on-state current is 15 A (max). It also has minimum gate trigger voltage of 2 V, minimum gate trigger current 100 mA and repetitive peak off-state voltage of 800 V (max).

Thyristors are shown in Figure 134. The inverter operation is such that only two thyristors, one in the top and the other in the bottom half of the bridge are fired at any given time by the application of a gate signal. If one considers thyristors 1 and 2 gated, the d.c. input voltage of the inverter will be impressed across the motor phase winding UV. When these two thyristors are off and thyristors 4 and 5 are on, the voltage across UV is reversed, thus generating

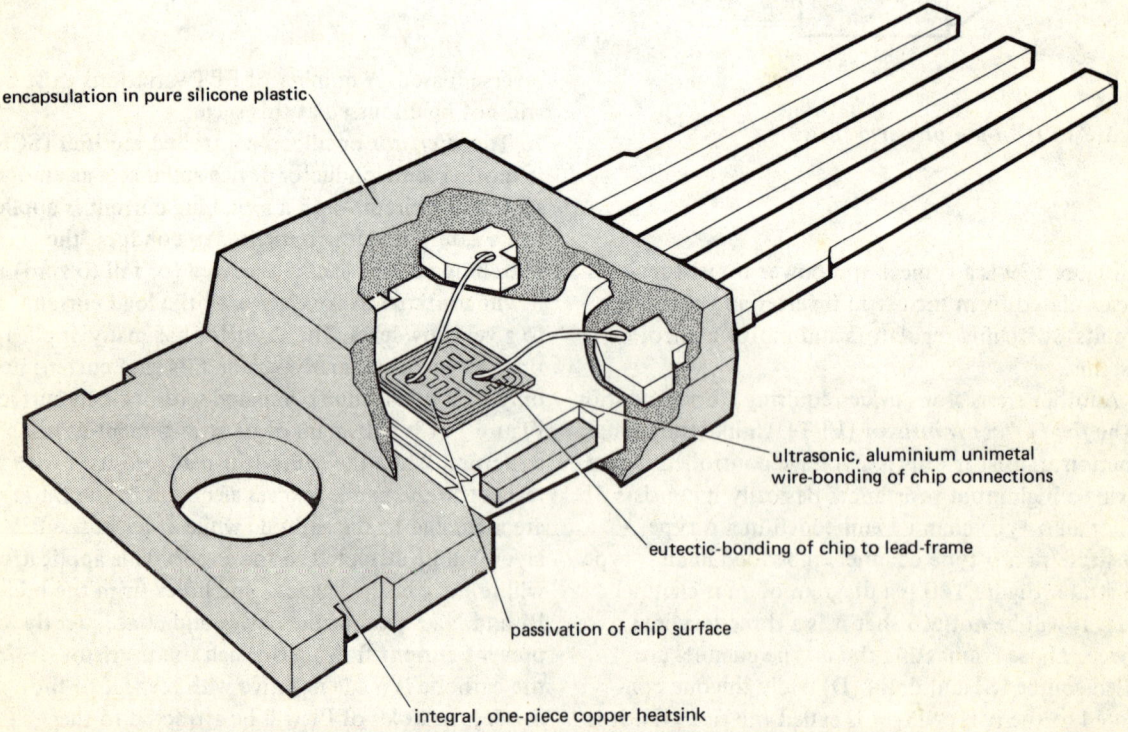

Figure 142   *Fast-turn-on thyristor assembly (TO-220 plastic, courtesy of Mullard)*

an alternating voltage. In application, thyristors 1, 3 and 5 are gated at 120° (elect.) intervals, as are thyristors 4, 6 and 2, so as to provide a three-phase waveform. It should be noted that the frequency of the a.c. voltage is controlled by the frequency at which the thyristors are fired. A typical industrial application for this method of control is in a re-heat furnace which uses roller conveyor motors.

The above description is only meant to provide an insight into the use of thyristors, the firing of their gates would take a little longer to explain. However, if one looks at the diagram of Figure 143 this might be more feasible. If one considers the thyristor's anode made negative with respect to the cathode it will not conduct, consequently, for one half of the supply cycle the motor receives no power. When the thyristor's anode is made positive with respect to the cathode, it will only conduct if the gate is suitably biased. Connected in the gate circuit is a diode which only conducts positive-going pulses. These pulses are controlled by $R2$; the more positive the gate is made, the larger will be the conduction angle and the higher the motor speed. While this sounds like a simple explanation it is not the full story since the motor will generate its own back e.m.f. which actually biases the thyristor's cathode positively. When the thyristor conducts the cathode potential takes up the anode value and D1 becomes reverse biased which in turn blocks the gate current. It is important therefore that this cathode bias is overcome which it will be by adjusting $R2$. In practice, there are a number of factors which may influence the firing circuits of thyristors, one of these is that the gate current must be maintained until the turn-on process is complete

and the thyristor is fully conducting; interference is sometimes another cause stemming from the power circuit itself or from relays and contactors in close proximity to the firing circuit. Figure 144 shows a

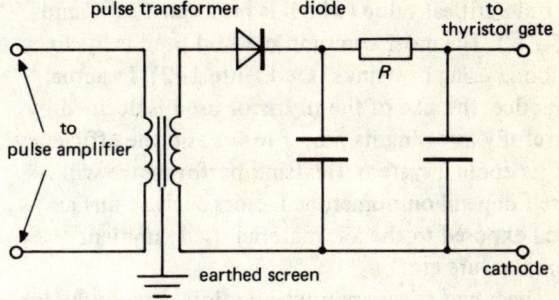

Figure 144   *Thyristor gate firing circuit*

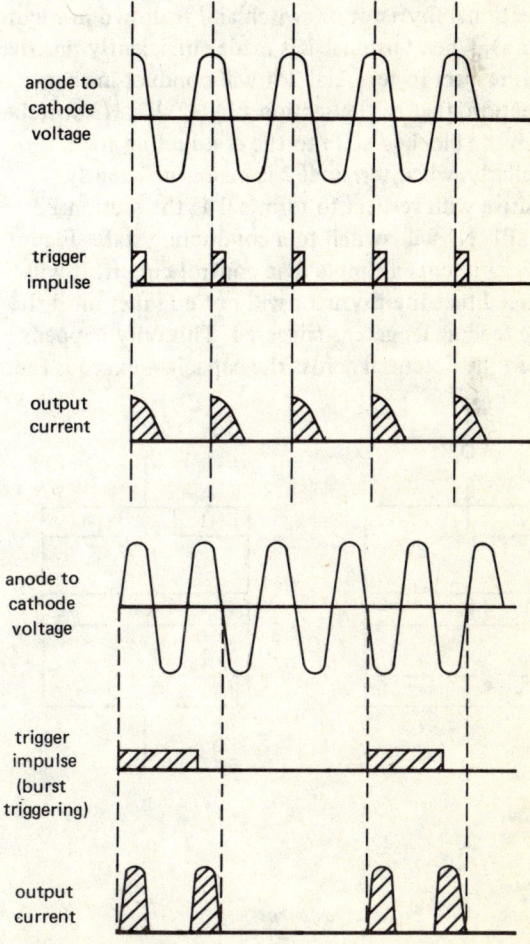

Figure 145   *Thyristor triggering circuits*

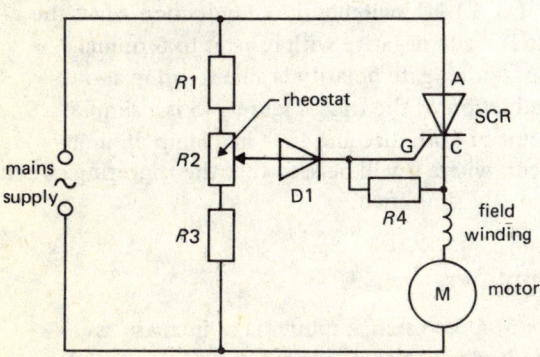

Figure 143   *Simple speed control of motor*

typical firing circuit while Figure 145 shows two output current traces for different trigger impulse lengths.

Because thyristors are solid-state devices, they need cooling. They will lose their blocking ability if the junction temperature is allowed to exceed a certain critical value (which is between 120°C and 150°C). The most common method used is by air cooling using heatsinks (see Figure 142). In actual practice, the size of the thyristor used is decided by carefully assessing its power losses and the efficiency of its cooling system. Heatsink performance will itself depend on numerous factors such as surface area exposed to the air, material used, ambient temperature etc.

Diacs and triacs are multijunction semiconductor devices which are triggered into conduction in either the forward or reverse direction. The diac is a bi-directional thyristor or switch and is shown in Figure 146(a). When terminal 1 is made sufficiently positive with respect to terminal 2 it will conduct in that direction, that is, the section P1, N2, P2, N3 switches from the blocked state to the conducting state. Similarly, when terminal 2 is made sufficiently positive with respect to terminal 1, the section P2, N2, P1, N1 will switch to a conducting state. Figure 147 represents a simple diac control circuit. It will be noticed that the thyristor will not conduct until the diac feeding its gate is triggered. This only happens when the potential across the capacitor exceeds the

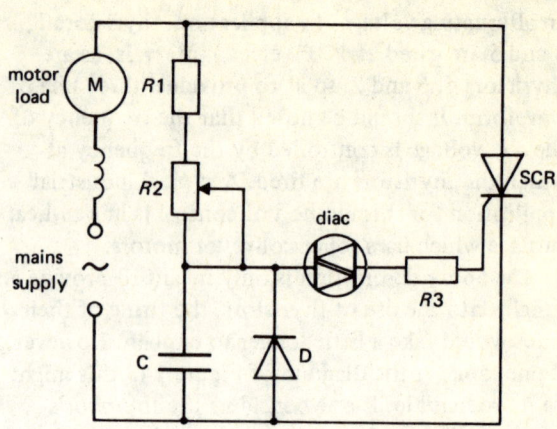

Figure 147   *Simple diac control circuit*

breakover voltage of the diac and the diac then conducts. Both *R*1 and *R*2 govern the rate at which the capacitor is charged but *R2* controls the triggering and conduction angle of the thyristor, allowing the load to have an increase or decrease in power. Resistor *R3* is connected in circuit to limit the gate current to a safe level and diode D ensures a stable triggering pulse by discharging during each negative half cycle of the supply. The circuit shown is suitable for speed control of small motors.

With regard to the triac in Figure 146(b), which is a gated bidirectional switch, when terminal 2 is made sufficiently positive with respect to terminal 1, the region P1, N3, P2, N5 acts like a thyristor which can be triggered into conduction by making the gate positive with respect to terminal 1. When this happens, the three junctions P1N1, P1N2 and P2N4 become reverse biased. However, when terminal 1 is made positive with respect to terminal 2, the region P2, N3, P1 N1 switches into conduction when the gate is made negative with respect to terminal 1. Thus, either gate polarity is able to bring about conduction of the triac. Figure 148 is a simple circuit of both diac and triac in a lamp dimming circuit where it will be seen that the triggering circuit is greatly simplified.

Amplifiers

The purpose of an amplifier is to increase the amplitude of voltage, current or power of an electrical circuit. If an amplifier is used to amplify either

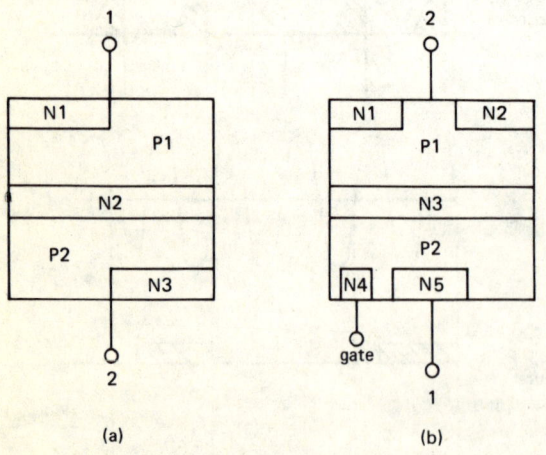

Figure 146   *Basic structures*
                (a) Diac
                (b) Triac

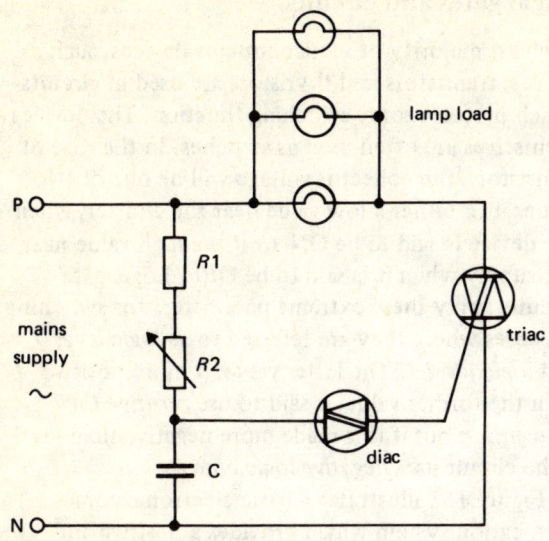

Figure 148   *Simple diac and triac lamp dimmer control circuit*

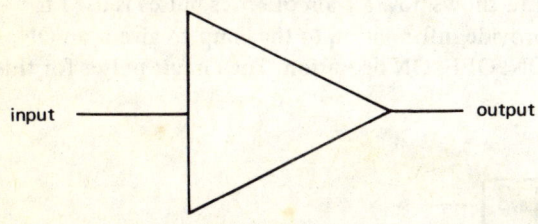

Figure 149   *BS 3939 symbol for an amplifier*

current or voltage such that its output signal is not distorted, it is called a *small signal amplifier* but if it is used to amplify the power of an input signal then it is called a *large signal amplifier*.

An amplifier has basically three properties, namely, *gain, frequency response* and a *transfer characteristic*. The gain is the amount of increase of its output signal amplitude compared with what it was before amplification, the frequency response is the way in which the gain varies with frequency and the transfer characteristic indicates the amount of gain that may be obtained from an amplifying device. Figure 149 is the symbol used for illustrating an amplifier.

Figure 150 shows three basic forms of connecting transistor amplifiers and while they do not show the base d.c. bias which they require, they plainly show the input and output connections. In diagram (a) the input is applied between the emitter and base while the output is generated between collector and base. In this way the connection of the base is common to input and output. The common base amplifier exhibits a *very low* input resistance and very high output resistance, providing *high* voltage gain and unity current gain which make it ideal for v.h.f. and u.h.f. receiver application. In diagram (b), the input is again applied between base and emitter but the output is generated between collector and emitter. The common emitter amplifier is the most popular one used since it has *low* input resistance and *high* out-

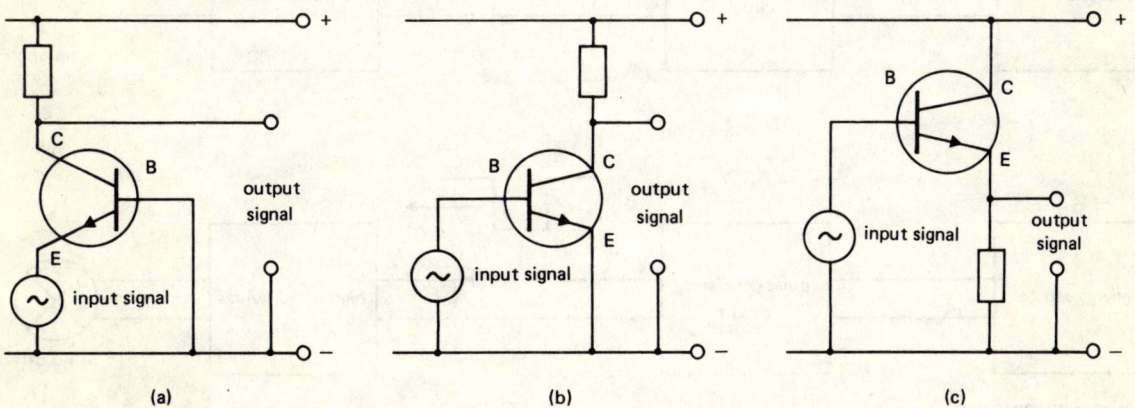

Figure 150   *Transistor amplifiers*
  (a) Common base
  (b) Common emitter
  (c) Common collector or sometimes called emitter follower

Note: The above are basic circuits with d.c. bias being omitted for simplicity

put resistance, giving *high* voltage gain and *high* current gain. It finds considerable use in oscillator and pulse generator circuits. Coming to diagram (c), the input connections are between base and collector while the output is generated between emitter and collector. This amplifier is often called an *emitter-follower* having a high input resistance and very low output resistance. It produces a high current gain and unity voltage gain and it is frequently used as a buffer amplifier to provide a high input resistance and as a final stage to give a low output resistance.

It should be pointed out that when a transistor is switched on, the power dissipated at the collector-base junction causes the junction's temperature to rise. This produces a *leakage current*, which although independent of the input current, can be the cause of breakdown of the device if not checked. The leakage current in a common-emitter amplifier can be quite considerable since its flow through the base–emitter junction to the supply terminals is amplified. The regenerative build-up of temperature and leakage current is called *thermal runaway* and is often checked by a stabilizing circuit and using a heatsink to limit the temperature rise.

Logic gates and circuits

The vast majority of semiconductor devices, such as diodes, transistors and thyristors are used in circuits which perform some switching function. The devices themselves are often used as switches. In the case of transistors, the collector voltage will be one of two values, it is either a low value near the emitter, when the device is said to be ON, or it is a high value near the supply when it is said to be OFF. Logic gate circuits apply these extreme parameters for switching purposes where they are referred to as *logic level 0* and *logic level 1*. The latter value, if more positive than the former value, is said to use *positive logic convention* but if it is made more negative than level 0 the circuit uses *negative logic convention*.

Figure 151 illustrates a basic electronic communication system which provides a positive and negative pulse via a high speed transistor switching arrangement. The positive pulse switches the lamp ON and the negative pulse switches it OFF. Figure 152 shows how a train of series pulses is used to provide information to the lamp to give it an ON, ON, OFF, ON operation. The circuit pulses for this

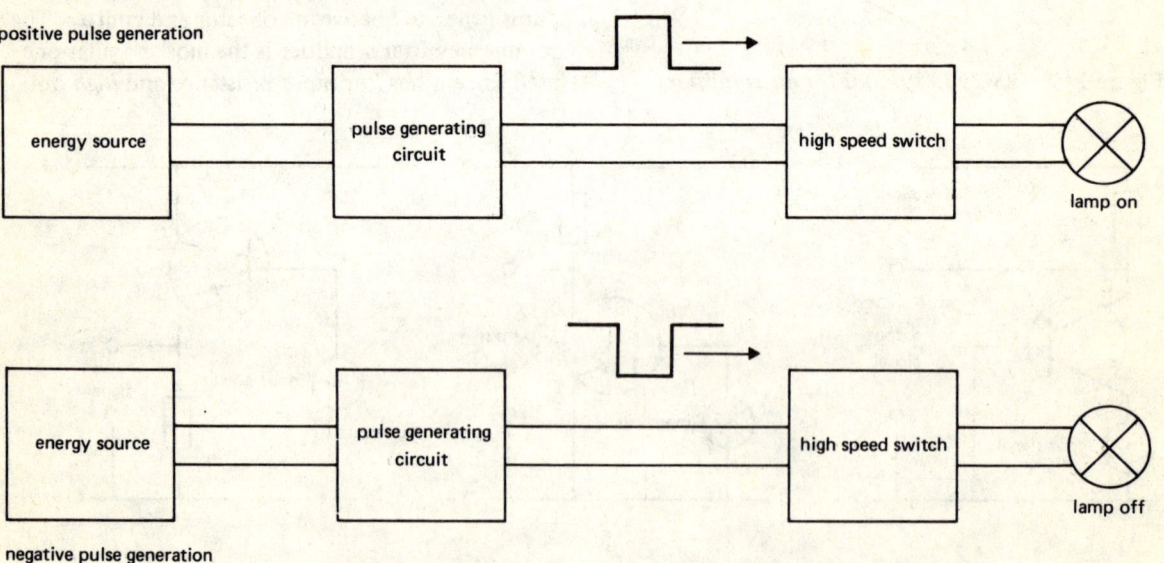

Figure 151   *Basic electronic communication system*

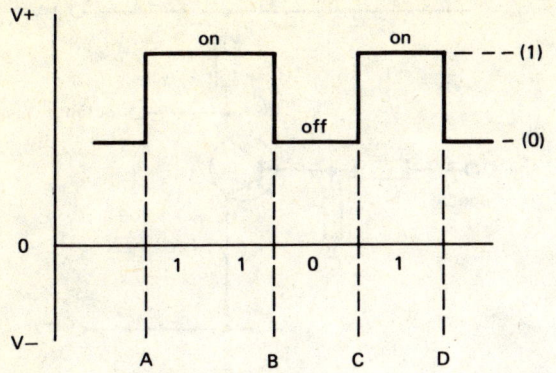

Figure 152   *Positive logic convention*

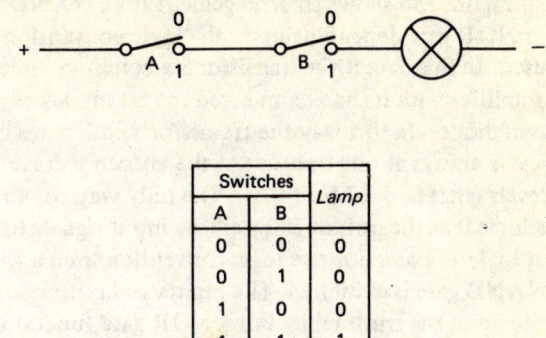

Figure 154   *Switching circuit and truth table for an AND gate*

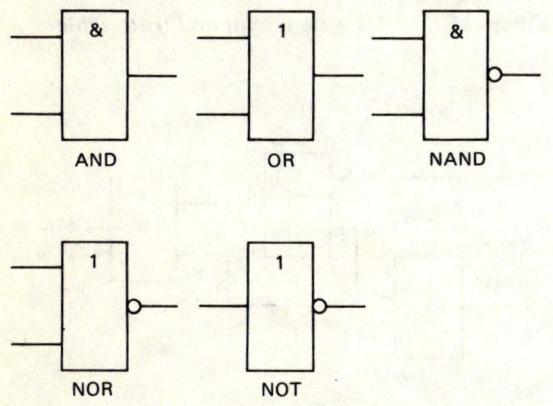

Figure 153   *Basic logic gate symbols*

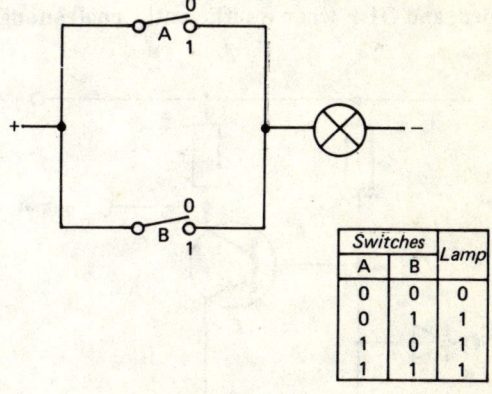

Figure 155   *Switching circuit and truth table for an OR gate*

are between points A and C where at D the lamp would be switched OFF. The circuit uses a positive logic convention.

The elements of a logic circuit can only have two values or states as they are sometimes called, these are ON or OFF, and it is the function of a *logic gate* to control the flow of information through the system utilizing these two conditions. Clearly, one is dealing with two forms of information which can be represented by a binary code in as much as the digits 1 and 0 can be used for the ON and OFF conditions, respectively. By doing this, mathematical operations are grossly simplified. The operation of logic gates

uses the expressions AND, OR, NAND, NOR and NOT. These gate symbols are shown in Figure 153.

The logical function of an AND gate is shown in Figure 154. It will be seen that the lamp will only light if A *and* B switches are both closed. This implies that in order to obtain a logic 1 output, all the inputs must be at logic 1. This is illustrated in the *truth table* at the side of the circuit.

The logical function of an OR gate is shown in Figure 155 using the two switches connected in parallel. The closing of A *or* B will cause the lamp to light. In both the AND and OR circuits the input signals are the same polarity as the output signals.

Figure 156 shows an arrangement for a NAND or NOR gate depending upon the logic convention used. In this circuit the transistor is a common-emitter amplifier with its base connected to the anodes of two diodes. In this way the transistor's collector voltage is always at one or other of the chosen voltage levels (that is, + 12 V or 0 V). The only way to obtain a logic 0 at the output is for all the input signals to be at logic 1. For a positive logic convention used a NAND gate is exhibited. The inputs and output are shown in the truth table. For a NOR gate function, the circuit has to be given negative logic convention such that when any of the input signals is at logic 1, the output is at logic 0. Conversely, to obtain a logic 1 at the output, the inputs need to be at logic 0.

Figure 157 illustrates a NOT gate circuit. The transistor is ON when sufficient base current flows, that is, when large enough positive potential is applied, and OFF when a sufficiently small enough

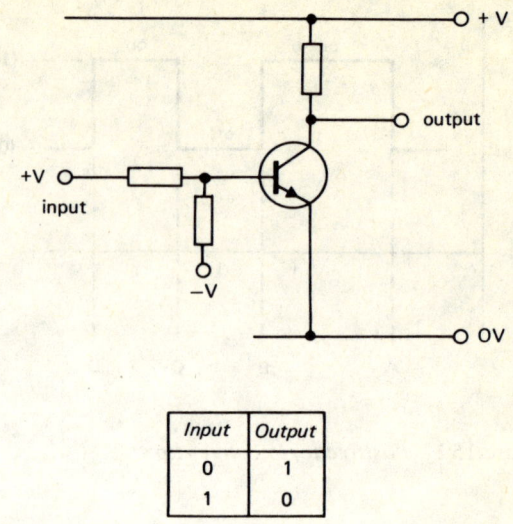

Input	Output
0	1
1	0

Figure 157   *NOT gate circuit and truth table*

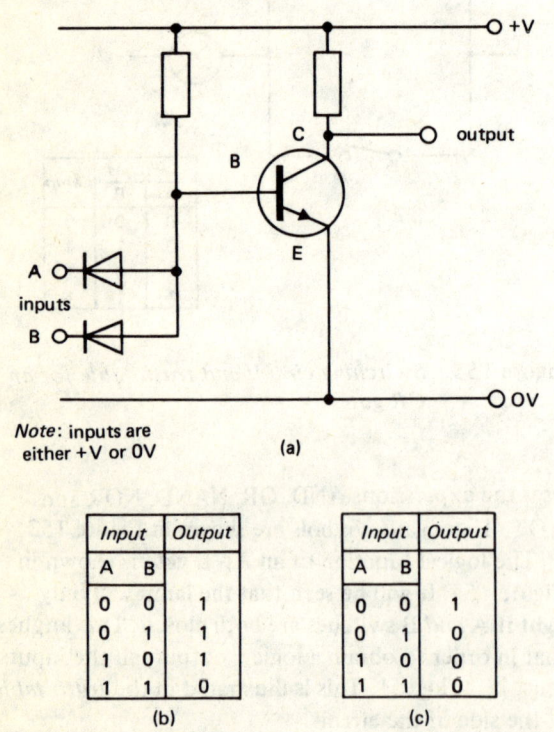

Note: inputs are either +V or 0V   (a)

Input		Output
A	B	
0	0	1
0	1	1
1	0	1
1	1	0

(b)

Input		Output
A	B	
0	0	1
0	1	0
1	0	0
1	1	0

(c)

Figure 156   (a) *Method of obtaining NAND and*
                     NOR gates
               (b) *NAND truth table*
               (c) *NOR truth table*

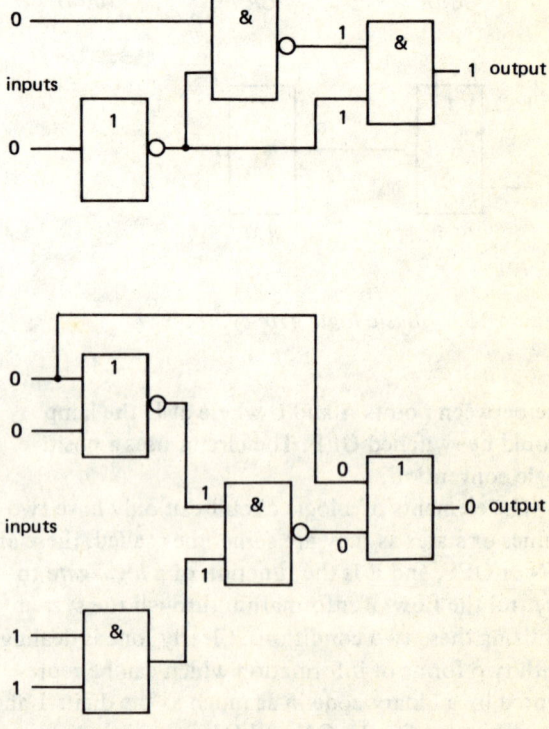

Figure 158   *Logic gates showing input and output*
                     functions

negative potential is applied. If positive logic convention is used, zero potential represents the 0 state and a positive potential represents a 1 state. Thus when no input pulses are applied the transistor is switched off and no current will flow through the collector. The output is therefore at a logic state 1. By applying pulses at the input (logic state 1), the transistor is switched ON and this has the effect of making the output potential zero volts which is logic state 0. The truth table for this function is also shown.

Figure 158 shows the connections of different logic gates with their respective inputs and output. The sequence uses the positive logic convention and information derived from the truth tables for each logic gate in circuit. Figure 159 shows three common integrated circuit (IC) packages which function on the above logic switching such as found in electronic calculators and adding machines. Figure 160 represents the logic gate switching for a machine motor contactor control. The output signal will be put through an amplifier circuit to actually energize the contactor. The circuit is such that the motor can only run if the machine guard is in position and the coolant pump is on. The motor will stop if there is an overload or an earth leakage fault. For maintenance and setting up the machine an override is provided so that the motor will run protected by the overload and earth leakage but the guard and coolant pump are off.

Revision exercise 6

1 With reference to Figure 161 label the components numbered 1 to 10.

2 Determine the magnitude and frequency of the trace shown in Figure 162 when the volt/cm control is set on 20 and the time base control is set on 1 μs.

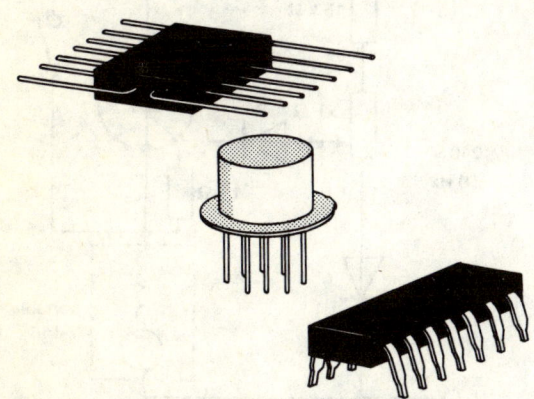

Figure 159   *IC packages*

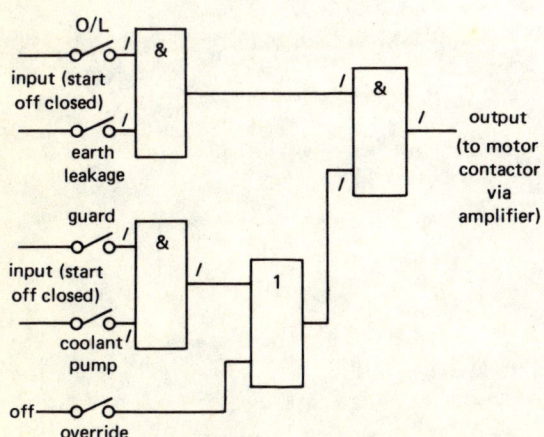

Figure 160   *Machine motor control*

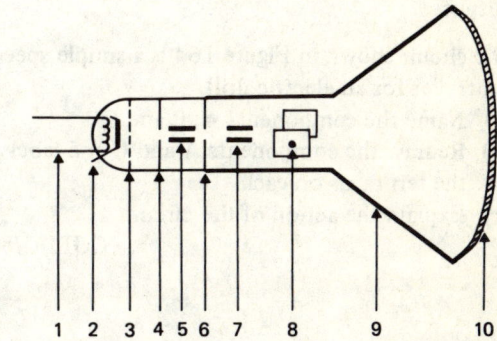

Figure 161

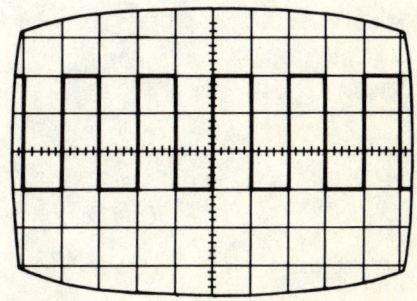

Figure 162

3 Explain with the aid of a diagram the operation
   of *one* of the following transducer devices:
   (a) piezoelectric transducer
   (b) thermoelectric transducer
   (c) variable resistance transducer

4 (a) Describe the action of a thyristor
   (b) Explain with the aid of a circuit diagram *one*
       typical application of such a device
                                          CGLI/C/82

5 Draw a diagram of a simple pnp phototransistor
   circuit, labelling its connections. Briefly explain
   its operation.

6 Give *two* reasons why *heatsinks* are used for
   semiconductor junction devices.

7 Distinguish between an AND and OR logic gate
   with the aid of a diagram.

8 Write a *truth* table for a NAND logic gate
   having *two* inputs and *one* output.

9 Assuming a positive logic convention, in Figure
   163, determine the input states for requiring a 0
   output.

10 The circuit shown in Figure 164 is a simple speed
   controller for an electric drill.
   (a) Name the components A, B and C
   (b) Redraw the components A and C and label
       the terminals on each
   (c) Explain the action of the circuit
                                          CGLI/C/82

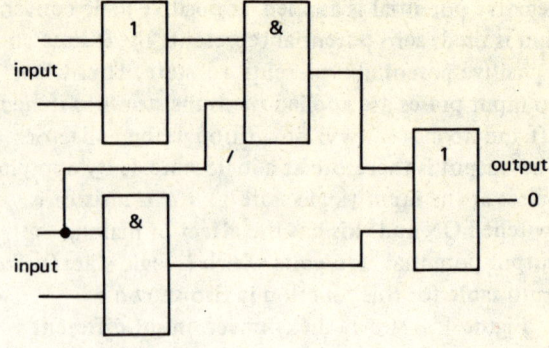

Figure 163

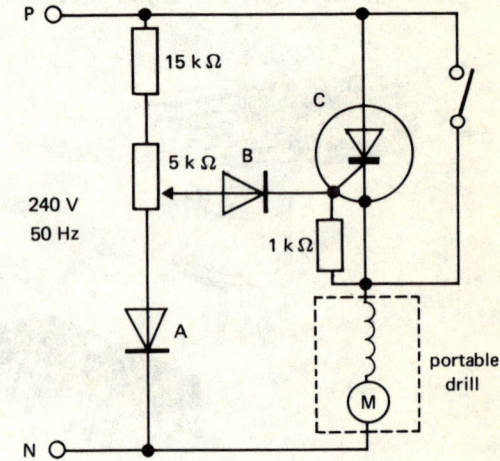

Figure 164

Answers to revision exercises

Revision exercise 1

1 The answers to this question can be found on pages 9–10.

2 (a) R_{ins} = 0.024 megohms.
   (b) A mains service cable will cater for three-phases of the supply, namely, red, yellow and blue phases (numbered 1, 2 and 3 respectively). Where neutral is used it is numbered 0.
   (c) Variation of supply voltage is between 390.1 V and 439.9 V whereas frequency variation is between 49.5 Hz and 50.5 Hz.

3 (a) Excessive voltage drop causes apparatus to work inefficiently ($P = V^2/R$); it also gives rise to unnecessary heat losses in final circuit wiring. See IEE Wiring Regulations, Page 522–8.
   (b) Prospective short-circuit current is the magnitude of the fault current likely to occur under extreme short-circuit conditions. Its determination is important in order to correctly install protective equipment capable of handling and clearing short-circuit faults when and where they arise. See IEE Wiring Regulations, Sections 462, 476 and 537.

4 (a) See page 21 of the NICEIC handbook on the 15th Edition of the IEE Wiring Regulations.
   (b) See (a) above and also IEE Wiring Regulations, Part 2 'Definitions' and Part 4 'Protection for safety'.

5 For any of the areas mentioned give *reasons*. For example, IEE Reg. 471–34 is one of six regulations concerning *bathrooms and showers* and points out that no provision shall be made for connecting portable equipment and no socket outlets are allowed in such rooms. Shaver sockets complying with BS 3052 are allowed because of their safety features (secondary winding is isolated from the supply) but the transformer core earthing terminal must be connected to the protective conductor of the final circuit.
   The regulation is an attempt to avoid the risk of electric shock from potentially dangerous equipment in a wet environment. It is pointed

out that socket outlets should be situated at least 2.5 m away from a shower cubicle, thus avoiding any splashing or spraying of water on to them. This is a relaxation only if the room is not a bathroom.

6　For further reference to these topics, see Appendix 1 of the IEE Wiring Regulations. Some notes on BS 5266 can be found in this book on pages 72-4.

7　(a)　Three tests:
　　　(i)　Test in accordance with item 6 of Appendix 15 of the IEE Wiring Regulations.
　　　(ii)　Test earth fault loop impedance. The RCD should be one rated for an operating current not exceeding 30 mA to protect against electric shock.
　　　(iii)　Test operation of the RCD test button to ascertain sticking of trip mechanism or faulty detector winding.
　　(b)　The premises is one likely to be damp and RCDs are much more reliable than voltage-operated devices in operation. They are also required to be used with socket outlets supplying portable equipment, particularly equipment used outside the equipotential earth zone. Selectivity and sensitivity are their two main advantages over fault voltage units. One major disadvantage with the RCD is that it cannot be used on a system with a high fault loop impedance or a system having a PEN conductor throughout.

Revision exercise 2

1　(a)　(i)　This is the magnitude of the fault current that is likely to flow under extreme short-circuit conditions, that is, negligible impedance between live conductors which are at different potentials under normal operating conditions.
　　　(ii)　Switchgear needs to be selected which is capable of handling short-circuit currents that would otherwise cause damage. When such conditions occur, overcurrent protective devices must provide rapid disconnection of the supply and their

breaking capacities need to exceed the prospective current level found in that part of the circuit they protect.
　　(b)　The answer to this part of the question can be found in Chapter 2. Briefly, resistance doesn't present a problem on the supply side since reactance of transformer windings and other apparatus having reactance provide the most opposition to short-circuit faults. Resistance of final circuit cables becomes more significant on the consumers' premises.
　　(c)　See Chapter 2.
　　(d)　Short-circuit MVA = 15.
　　　　Short-circuit current = 787 A.

2　(a)　2.6%.
　　(b)　440.55 A.

3　(a)　See Chapter 2.
　　(b)　See Chapter 2.
　　(c)　Buchholz relay.

4　(a)　See cable manufacturers' catalogues.
　　(b)　Hints:
　　　(i)　Measure cable to point of entry, mark off and allow sufficient tails.
　　　(ii)　Strip cable for termination taking necessary precautions.
　　　(iii)　Fit and tighten cable gland if PVC cable or plumb joint lead cable – terminate conductors.
　　　(iv)　Identify conductors where required.
　　　(v)　Provide adequate earthing and test cable.
　　(c)　Precautions include:
　　　(i)　Avoid damage to cable insulation.
　　　(ii)　Use cable jacks and rollers.
　　　(iii)　Use sand base in trench.
　　　(iv)　Remove large stones from trench.
　　　(v)　Use non-corrosive materials and provide further protection of the cable where necessary.

5　50 mm^2 – if HBC fuses protect cable.
　70 mm^2 – if semi-enclosed fuses protect cable.

6　(a)　(i)　Conductor operating temperature is higher.
　　　(ii)　Conductor terminations are easier.
　　　(iii)　Installation time is less.
　　(b)　See manufacturers' literature.

7 (a) See IEE Wiring Regulations, Part 2
        'Definitions'. See also Regs. 413–2, 413–7
        and Section 547.
   (b) See IEE Regs. 471–34 to 471–39.
   (c) See BS 3052. The shaver unit contains a
        double wound transformer with its core
        earthed to provide protection against leakage
        between the primary and secondary
        windings. The two windings are isolated
        from each other with the secondary winding
        also providing 110 V.

8 See CP 1017, BS 4343 and BS 4363.

Revision exercise 3

1 See Chapter 3.

2 See Chapter 3.

3 See Chapter 3.

4 See Chapter 3.

5 Hints:
   separate answers into wiring systems, switchgear
   and protective devices; write notes on each type
   of final circuit. See also the *Farm Electric* hand-
   books published by the Electricity Council.

6 (a) See Chapter 3, but think about presence of
        livestock, water, corrosive atmosphere,
        mechanical damage and neglect.
   (b) Cowsheds: mechanical protection required;
        wiring system needs to consider corrosive
        atmosphere, for example ammonia.
        Grain drying areas: dusty atmosphere
        (perhaps).
        Storage area for heavy machinery; damage
        to wiring system.

7 See Chapter 3.

8 See Chapter 3.

9 See Chapter 3.

10 Draw up a list with headings such as 'quantity'
    and 'description' and consult manufacturers'
    catalogues.

11 See Chapter 3.

12 See Chapter 3.

13 (a) Inbalance of electrons in a material.
    (b) Movement of rubber conveyor belts and
         build up of charge in electronic assembly
         work. See also text regarding hospital
         operating theatres on page 50.
    (c) Static charge could result in ignition and
         explosion. The conveyor belt may release a
         discharge into a hostile environment contain-
         ing explosive liquids and gases. The danger
         can be minimized by installing static dis-
         charge brushes which make contact with the
         conveyor. With regard to electronic work,
         special anti-static components and materials
         should be used and personnel should not
         wear nylon clothing.

Revision exercise 4

1 See Chapter 4. See also Chapter 6 of Volume 2
   of this series.

2 Make reference to 'High voltage discharge
   lighting', IEE Regs. 554–3 to 554–19.
   (a) Sketch should show termination of HV cable
        (often lead sheath cable) into a bell glass or
        glazed porcelain cover. Measurements of
        support spacing should be inserted.
   (b) Sketch should show lead sheath cable and
        packing gland connected to transformer case
        with bonding arrangement shown where
        required.
   (c) Sketch should show tube attached to metal
        letter with lock-nutted stud, phosphor
        bronze clip and binding wire.

3 See Chapter 4.

4 See Chapter 4.

5 (a) See Chapter 4.
   (b) See Chapter 3 of Volume 2 for control of
        supply for off-peak tariff.

6 (a) Immediate availability of hot water. Auto-
        matic cut-off eliminates wastage of water
        and cuts electricity costs too. Unit may be
        provided with selected temperature control.
   (b) 6.3 kW.
   (c) Instant shower costs are 1.57p for 3 minutes
        whereas hot water system costs are 15p to
        fill bath.

7   Find the necessary power required to heat the air, remembering that 1 kWh = 3.6 MJ. This is 6.26 kW.

Find the necessary power losses from the room, remembering to add together the separate losses. This is 10.4 kW.

Total power required is 16.66 kW.

Revision exercise 5

1   See Chapter 5. See also Chapter 2 of Volume 2.

2   (a) Draw network carefully, remembering which activities follow each other. Note that activity C does not restrict any others, the bar chart being somewhat misleading.
    (b) The critical path is along BDH.
    (c) Total float on activity E is the earliest start time subtracted from the latest finish time, that is, $45 - 20 = 25$ weeks. The length of time that can be spent on this activity without increasing the total project time is $25 - 10 = 15$ weeks.

3   (a) Document kept in a site office.
    (b) The use of a site diary is for recording information.
    (c) Telephone calls, visits, correspondence, site meetings, etc.

4   (a) In the lump sum contract, the electrical contractor takes off quantities which are submitted by the client and any errors made in the contractor's calculations cannot be requited.

With regard to the bill of quantities contract, the electrical contractor's tender price is made up from a number of unit price items and the contractor will only be paid for the number of items installed.
    (b) See Chapter 5.

5   See Chapter 5.

6   Project size, location and contact document. The first may be considered too large for the contractor to undertake, the second may present problems of supervision if too far away from the general office, and the third may contain clauses that are unacceptable. Cash flow and labour resources are other factors.

7   See the JIB National Working Rules and EETPU literature.

8   (a) 0-1, 1-8, 8-9, 9-10, 10-12, 12-13. Draw network.
    (b) The activities on the critical path represent the longest time taken to complete the project, that is, 49 days. The effect of activity 6-7 taking 7 days longer alters the critical path to activities 0-1, 1-5, 6-7 and 7-13 taking 50 days.

Revision exercise 6

The majority of questions in this exercise can be answered by reading Chapter 6.

Appendix 1
Estimating
material

CONTRACT CONDITIONS CHECK LIST

Contract Name ....   Date ....

Enquiry No. ....   Estimator Ref..........................

Architect...............................   Quantity Surveyor.......................

Consultant..............................   Main Contractor ....

Tick appropriate boxes and complete spaces below to indicate clearly what the tender documents require XYZ to adhere to.
Add notes where appropriate to show where XYZ tender must be qualified.

TYPE OF CONTRACT

R.I.B.A. Nominated (Green)   ☑   Contract edition Revision Date ....

        Direct (Blue)   ☐   Fixed Price   ☐   Variable Price   ☑

        GC/Works/1   ☐   Bill of Quantities ☐   XYZ Take-off   ☑

Other.....................................

PROGRAMME

Main Contract Dates .AUG. '83 - JUNE '84   XYZ Contract Dates .MAR. '83 - JUNE '84

                   or/   To be agreed   ☐

                   or/   As and when directed ☐

LIQUIDATED DAMAGES

Amount per week £. 5,000 ....   23j (i)   in   ☑   out   ☐

                              (ii) in   ☑   out   ☐

CASH FLOW CONDITIONS

Payment   14 days after Architects
to be   Certificate   ☑   Retention on claim .... 5 %
made
        Pay when paid ✳   ☐   Limit of Retention .... 2½ %

        Other.................   % Release on Practical Completion ...... 50 %

Defects Liability Period...................   % Held during defects period. 1¼ OF LIMIT ... %

INSURANCE REQUIREMENTS

                                   Any additions or
                                   alterations to
Third Party   ☐   R.I.B.A. Third Party ....

Contract Works   ☑

INCREASED COSTS

R.I.B.A. Full ACD   ☐   Labour increases from:

        Part BCD   ☑   Date of Tender   ☐

Formula basis   ☐   Date of Tender
                                   but include
Percentage addition...................   promulgated increases   ☐

              ✳ DIRECTORS APPROVAL REQUIRED
                 FOR PAY WHEN PAY CONDITION

CONTRACT SUPERVISION

Resident   Full time
Site Staff   resident   ☐   As required by XYZ   ☑

GENERAL SITE RESPONSIBILITIES

	By Builder	By XYZ
Scaffolding as and when erected	[x]	[ ]
Special scaffolding	[ ]	[x]
Hoisting and positioning	[x]	[ ]
Protection of works		
installed but not handed over	[ ]	[x]
Storage facilities	[ ]	[ ]

	By Builder	By XYZ
Site power supply		
415v	[ ]	[ ]
240v	[ ]	[ ]
110v	[x]	[ ]
To offices	[ ]	[ ]
Site temporary lighting supply	[x]	[ ]
Site water supply	[x]	[ ]
To offices	[ ]	[ ]
Off-loading	[x]	[ ]
Health and welfare facilities	[x]	[ ]

CO-ORDINATION AND WORKING DRAWINGS

Consultants Drawings	Plant Room	Other Areas
1/200	[x]	[ ]
1/100	[ ]	[x]
1/50	[ ]	[x]
1/25	[ ]	[ ]

Co-ordination of other sub-contractors services

By Builder	[x]
By XYZ/Builder	[ ]
By XYZ	[ ]

Any special drawing requirement . . . .

DESIGN RESPONSIBILITIES

XYZ total design responsibility [ ]

XYZ no design responsibility [x]

XYZ part design responsibility

Acoustics	[ ]
Controls	[ ]
Pump/fan size checks	[ ]
Other	

ANY OTHER SPECIAL CONDITIONS

SECTION SUMMARY SHEET

Contract Name | Town | Enquiry No. | Estimator | Date

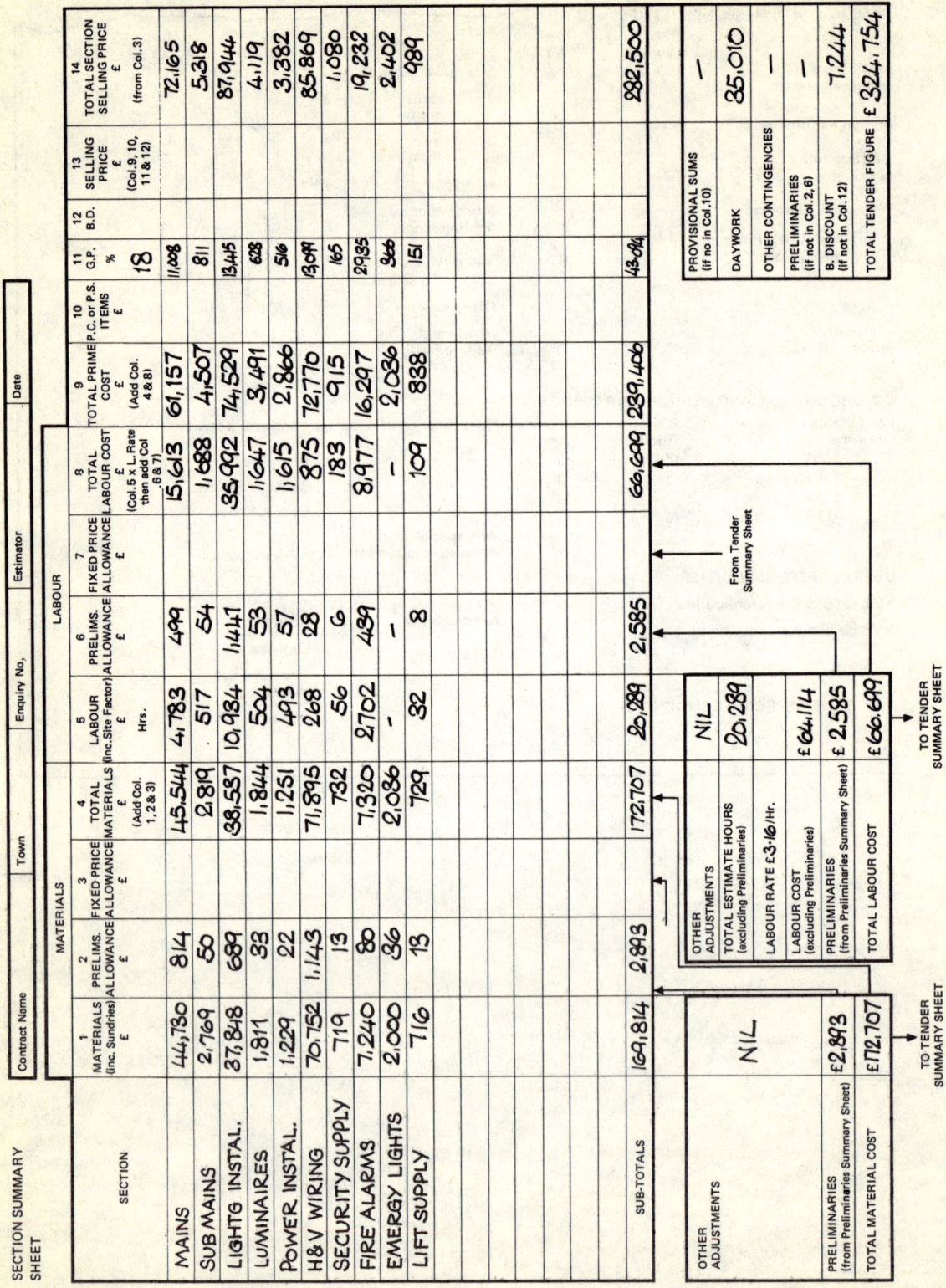

SECTION	MATERIALS				LABOUR				9 TOTAL PRIME COST (Add Col. 4 & 8) £	10 P.C. or P.S. ITEMS £	11 G.P. %	12 B.D.	13 SELLING PRICE (Col.9,10,11 & 12) £	14 TOTAL SECTION SELLING PRICE (from Col.3) £
	1 MATERIALS (inc. Sundries) £	2 PRELIMS. ALLOWANCE £	3 FIXED PRICE ALLOWANCE £	4 TOTAL MATERIALS (Add Col. 1,2 & 3) £	5 LABOUR (inc. Site Factor) Hrs.	6 PRELIMS. ALLOWANCE £	7 FIXED PRICE ALLOWANCE £	8 TOTAL LABOUR COST (Col.5 × L.Rate then add Col .6 & 7) £						
MAINS	44,730	814		45,544	4,783	499		15,613	61,157		11,008			72,165
SUB MAINS	2,769	50		2,819	517	54		1,688	4,507		811			5,318
LIGHTG INSTAL.	37,848	689		38,537	10,934	1,441		35,992	74,529		13,415			87,944
LUMINAIRES	1,811	33		1,844	504	53		1,647	3,491		628			4,119
POWER INSTAL.	1,229	22		1,251	493	57		1,615	2,866		516			3,382
H & V WIRING	70,752	1,143		71,895	268	28		875	72,770		13,099			85,869
SECURITY SUPPLY	719	13		732	56	6		183	915		165			1,080
FIRE ALARMS	7,240	80		7,320	2702	439		8,977	16,297		2935			19,232
EMERGY LIGHTS	2,000	36		2,036	-	-		-	2,036		366			2,402
LIFT SUPPLY	716	13		729	32	8		109	838		151			989
SUB-TOTALS	169,814	2,893		172,707	20,289	2,585		66,699	239,406		43,094			282,500

(Col. 11 rate: 18)

OTHER ADJUSTMENTS — NIL

PRELIMINARIES (from Preliminaries Summary Sheet) — £2,893
TOTAL MATERIAL COST — £172,707
TO TENDER SUMMARY SHEET

OTHER ADJUSTMENTS — NIL

TOTAL ESTIMATE HOURS (excluding Preliminaries) — 20,289
LABOUR RATE £3·16/Hr.
LABOUR COST (excluding Preliminaries) — £64,114
PRELIMINARIES (from Preliminaries Summary Sheet) — £2,585
TOTAL LABOUR COST — £66,699
TO TENDER SUMMARY SHEET

From Tender Summary Sheet

PROVISIONAL SUMS (if no in Col.10)	—
DAYWORK	35,010
OTHER CONTINGENCIES	—
PRELIMINARIES (if not in Col.2, 6)	—
B. DISCOUNT (if not in Col.12)	7,244
TOTAL TENDER FIGURE	**£ 324,754**

Note: This is a standard summary sheet reduced for example purposes

CALCULATION OF LABOUR COST/MAN HOUR

Contract Name _____ Estimator Ref. _____

Town _____ Date _____

Enquiry No. _____

Distance from Office			8				Miles	
	Mon.	Tue.	Wed.	Thu.	Fri.	Sat.	Sun.	

	Mon.	Tue.	Wed.	Thu.	Fri.	Sat.	Sun.			
Hrs. Worked /Week	10	10	10	10	10	8		58	A
Hrs. Paid /Week			54			13½		67½	B

Contract Period	104 Wks.	
Period	82 Wks.	C
Total Estimate Man Hrs. (including Preliminaries)	26,284 Hrs.	D
Average Productive Men/Week $\frac{D}{A \times C}$	5 Men	E

BASIC WAGES/MAN/WEEK (Inc. N.P.T.) } SEE REVERSE SIDE FOR CALCULATION → £ 116·10 F

FARES, ALLOWANCES AND EXPENSES/MAN/WEEK _____ → £ 19·50 H

MENS RESERVE/MAN/WEEK _____ → £ 29·65 G

COST/MAN HOUR AT CURRENT RATES OF PAY $= \frac{F+G+H}{A} = \frac{105·25}{58}$ 2·85

Add _____ % for   —

Add _____ % for   —

—

ESTIMATES LABOUR COST PER MAN HOUR FOR CONTRACT _____ → £ 2·85 J

PRELIMINARIES SUMMARY SHEET

	Labour Time Hrs	Material Cost £
SETTING UP SITE		
Site Huts — (Offices, Stores and Messing)	45	100
Furniture	—	—
Signboard	—	40
Installation of Telephone, Electrics and Heating	—	250
Burglar alarm and/or Floodlight	—	—
Tube rack	8	—
Vanning and Lorry charges	—	—
Others	—	—
OPERATING SITE		
Towers, Ladders and Steps	—	} 250
Vices, Benches, Stocks and Dies	—	
Expendable Tools	—	
Special Scaffolding	—	
Special Workshop equipment	—	
Lifting Tackle/Craneage	—	
Running cost of Telephone, Electrics and Heating 24 MTHS. x 20	—	480
Rates	—	
Special Insurances	—	
Salaried Site Staff Costs and Expenses 24 MTHS. x 40	—	960
Others	—	
COMPLETING SITE		
Instruments and Special Test Equipment	—	} 250
Spares for Testing/Client		
Hydraulic Testing	IN PRIME COST	
Commissioning and Testing installation	} 360	
Commissioning and Testing special plant		
Attendance on other Sub-Contractors	135	
Fuel or Power for Testing	—	
Record Drawings/Fuse Charts/Labels	} INC. IN PRIME COSTS	
Operating Manuals and Instructions		
Temporary services (Attendance and Insurance)		
Training Clients Staff	45	
Operating plant for Client	45	
Clearing excess materials and rubbish 25 x 10 YD LORRIES AT £22.50	180	562·50
Others	—	

		Labour Time Hrs	Material Cost £
	Total Labour Hours	818	
✳	Labour Rate £ 3·16 /Hr (From Labour Cost per Hour Sheet)		
	Labour Cost	£ 2,584·88	

SITE SUPERVISION
Non-productive
hourly paid
costs and expenses

	Non-working		£	
		C/H	—	
		Foreman	—	
		Supervisor	—	
		Storeman etc.	—	

	Labour Cost	Material Cost
TOTAL PRELIMINARIES COSTS (To Section Summary Sheet)	£ 2,584·88	£ 2,892·50

✳ *Includes other items*

TENDER SUMMARY SHEET

CONTRACT NAME		For Computer Use only
TOWN		
ENQUIRY NUMBER		
BRANCH		
ARCHITECT		
CONSULTANT		
QUANTITY SURVEYOR		
CLIENT/BDG OWNER		

DATE

Enter One Number Only in Box Shown

1. Original Tender	2. Revision to Previous tender	1
1. Design	2. Non-Design	2
1. Fixed	2. Variable	2
SERVICES	1 M (inc.P) / 2 E / 3 M&E / 4 P only	1
MARKETING CODE	Competition	
	Building Type	

% Prime Cost		
%		
%		
18 %		

MATERIALS   £ 172,707

    Fixed Price   £ —   £172,707

LABOUR   £ 66,699

    Fixed Price   £ —   £ 66,699

TOTAL PRIME COST   [X] →   £ 239,406

OVERHEADS   £

PROFIT   £

GROSS PROFITABILITY   [Y] →   £ 43,094

RECOVERY ON INCREASED COSTS   £ —

NETT SALES VALUE   [Z] →   £ 282,500

BUILDERS DISCOUNT 1/39th   £ 7,244

    £ 289,744

P.C. SUMS
    Add for profit and B. Discount   £   £ —

PROVISIONAL SUMS   DAYWORKS   £ 35,010

TENDER FIGURE   £ 324,754

Remarks   PROGRAMME 100 WKS.

ESTIMATING LABOUR COSTS

2.6 ELECTRICAL (ALL OFFICES)

2.6.1 Town Jobs (within 25 miles from office)

(a) WAGES

Grade	Basic Hourly Rate	
	London	*Provincial*
Technician	£2.13	£2.04
Approved Electrician	£1.84	£1.75
Electrician	£1.69	£1.60
Labourer	£1.33	£1.24
Apprentice 20 years old	£1.35	£1.28
Apprentice 19 years old	£1.10	£1.04
Apprentice 18 years old	.85	.80
Apprentice 17 years old	.77	.72
Apprentice 16 years old	.68	.64

The above rates of pay take into account a working week of 38 hours, but it must be rembered that 42 hours have to be worked in any week at normal rates before overtime premium is calculated.

The above rates of pay *do not* include for any responsibility monies that may be paid to operatives.

(b) FARES, ALLOWANCES AND EXPENSES

Fares and expenses should be individually assessed for each contract using head office as base.

Travelling time should be included as follows:—

Distance *(point to point)*	Total Daily Travelling Time *(ordinary rates)* *Per day plus Actual Fares*
Return journey of up to	
1 mile each way	Nil
2 miles each way	20 minutes
3 miles each way	30 minutes
4 miles each way	40 minutes
5 miles each way	50 minutes
5 – 10 miles each way	75 minutes
10 – 15 miles each way	100 minutes
15 – 20 miles each way	110 minutes
20 – 25 miles each way	120 minutes

(c) MENS RESERVE

The mens reserve allowance is an ancillary cost which must be added to an estimate to allow for items mentioned in paragraph 2.2.1 (c)

	London	Provincial
Mens Reserve Cost	£29.65	£28.88

6.2 COUNTRY JOBS (OVER 25 MILES FROM OFFICE)

(a) WAGES

All as Electrical Town Jobs — see paragraph 7.6.1 (a)

(b) FARES, ALLOWANCES AND EXPENSES

(i) For an operative travelling each day to the contract £4.50 man/day

(ii) For an operative who is lodging away from home £31.50 man/week plus return fare —

Contracts up to 100 miles	— 1 return fare every two weeks
Contracts 100 miles up to 250 miles	— 1 return fare every four weeks
Contracts 250 miles upwards	— 1 return fare every four weeks plus 8 hours travelling time at basic hourly rate

Plus daily fares and travelling allowance where the distance between lodgings and the job involve more than ½ hour travelling each way and cost more than 10p/day in actual fares.

(c) MENS RESERVE

All as Electrical Town Jobs — see paragraph 2.4.1 (c)

CALCULATION OF AVERAGE NUMBER OF
PRODUCTIVE MEN ON SITE/ WEEK

Average No. of Productive men on site/week $=$ $\dfrac{\text{Total Estimate Man Hrs.}}{\text{Hrs.Wkd./Week} \times \text{MJN Weeks}} =$

$$= \frac{D}{A \times C} = \frac{26,284}{58 \times 82} = \boxed{5 \quad \text{men}} \; \boxed{E}$$

CALCULATION OF TOTAL BASIC WAGES/WEEK/MAN
FOR ELECTRICAL OPERATIVES

Grade	No. of men in Grade	Hrs.Paid/Wk. B	Rate of pay/Hr.	Total Pay £/Week
A/ ELEC.	1	$67\frac{1}{2} \times$	$1 \cdot 84 \times 1$	$124 \cdot 20$
ELEC.	4	$67\frac{1}{2} \times$	$1 \cdot 69 \times 4$	$456 \cdot 30$
Total Av. Prod. Men/Wk. **5**			Total Basic Wages/Wk.	£ $580 \cdot 50$ L

$$\frac{\text{Total Basic Wages/Week}}{\text{Av. Prod. Men/Week}} = \frac{L}{E} = \frac{580 \cdot 50}{5} = £\,116 \cdot 10$$

OTHER CALCULATIONS

TRAVELLING ALLOWANCE   SITE 8 MILES FROM OFFICE

75 MINUTES / MAN

AVERAGE RATE $\dfrac{L}{B \times E}$ $\dfrac{580 \cdot 5}{67\frac{1}{2} \times 5} = £1 \cdot 72 / \text{HOUR}.$ $£1 \cdot 72 / \text{HOUR} \times 1\frac{1}{4} \text{ HOURS} \times 6 \text{ DAYS} =$

$£12 \cdot 90$ PER WEEK.

FARES

$1 \cdot 10$ DAY RETURN × 6 DAYS $= £\dfrac{6 \cdot 60}{19 \cdot 50}$ PER WEEK.

Appendix 2
15th edition
IEE Wiring
Regulations

Terminology

Symbol   Meaning

U_O Nominal voltage (e.g. 240 V or 415 V)

I_p Prospective short circuit current (e.g. 16 kA)

I_B Design current of circuit (i.e. the load current)

I_N Nominal current or current setting of circuit protective device (e.g. BS 88, 32-A fuse or 6 A MCB)

I_Z Current carrying capacity of circuit conductor

I_f Each fault current (e.g. $I_f = \dfrac{U_O}{Z_S}$)

I_T Tabulated single-circuit current carrying capacity (e.g. $I_T = \dfrac{I_N}{F}$)

I_2 Effective operating current of device

F Group rating factor

CF Correction factor (e.g. C1 is a grouping factor, C2 is an ambient temperature factor, C3 is a thermal insulation factor and C4 is a fusing factor for BS 3036 semi-enclosed fuses, i.e. 0.725)

Z_t Impedance of the supply transformer

Z_e Earth fault loop impedance external to installation circuits. For a TNC-S system (PME) it is estimated to be 0.35 ohms, for a TNC system (concentric) it is estimated to be 0.35 ohms, for a TNS system (cable sheath or overhead protective conductor) it is estimated to be 0.8 ohms, and for a TT system (consumer's earth electrode) it is estimated to be 21 ohms.

Z_S Earth fault loop impedance. (For max. values see Tables 41A1 & 41A2 IEE Regs. For actual value use formula $Z_S = Z_e + R_1 + R_2$ Appendix 8, IEE Regs.)
Note $R_1 + R_2$ are found in Appendix 17, Table 17A, IEE Regs.

R_1 Resistance of phase conductor

R_2 Resistance of protective conductor

S Cross sectional area of protective conductor in mm² found by using the adiabatic equation, e.g.

$$S = \frac{\sqrt{I^2 t}}{k} \text{mm}^2$$

I is the fault current, t is the duration of the fault current or operating time of the protective device, and k is a factor for specific protective conductors (see Table 54B, IEE Regs.).

Further information

Examples of short circuit ratings of five common protective devices:

BS 3036 Re-wirable (semi-enclosed) fuse	1-4000 A
BS 1361 Domestic cartridge fuse	16000 A
BS 1362 Plug-top fuse	6000 A
BS 88 Part 2 h.b.c. cartridge fuse	80000 A
BS 3871 Types 1-3 MCBs	1-9000 A

Note   Residual current devices (RCDs) have a short circuit rating around 2000 A.

Cable selection procedure

1   Determine design current (I_B)

2   Determine protective device nominal current (I_n)

3   Apply correction factors (if any)

4   Determine cable size (I_Z)

$$I_Z = \frac{I_n}{C1 \times C2 \times C3 \times C4}$$

Note $I_B \leqslant I_n \leqslant I_Z$

5   Check voltage drop constraints (see IEE Regs. Reg. 522-8)

6   Check disconnection time of protective device

7   Check thermal constraints of protective conductor

Index